Information Circular 9508

The Application of Major Hazard Risk Assessment (MHRA) to Eliminate Multiple Fatality Occurrences in the US Minerals Industry

By A. Iannacchione, F. Varley and T. Brady

DEPARTMENT OF HEALTH AND HUMAN SERVICES
Centers for Disease Control and Prevention
National Institute for Occupational Safety and Health
Spokane Research Laboratory
Spokane, WA

October 2008

This document is in the public domain and may be freely copied or reprinted.

Disclaimer

Mention of any company or product does not constitute endorsement by the National Institute for Occupational Safety and Health (NIOSH). In addition, citations to Web sites external to NIOSH do not constitute NIOSH endorsement of the sponsoring organizations or their programs or products. Furthermore, NIOSH is not responsible for the content of these Web sites.

Ordering Information

To receive documents or other information about occupational safety and health topics, contact NIOSH at

>Telephone: **1–800–CDC–INFO** (1–800–232–4636)
>TTY: 1–888–232–6348
>e-mail: cdcinfo@cdc.gov

>or visit the NIOSH Web site at www.cdc.gov/niosh.

For a monthly update on news at NIOSH, subscribe to NIOSH eNews by visiting **www.cdc.gov/niosh/eNews**.

DHHS (NIOSH) Publication No. 2009–104

October 2008

SAFER • HEALTHIER • PEOPLE™

Table of Contents

Abstract 1
Executive Summary 2
Acknowledgement 3
1.0 - Introduction 3
   1.1 – Trends in Managing Major Mining Hazards 4
2.0 – Minerals Industry Risk Management 5
   2.1 – The US Experience 5
   2.2 – The Australian Experience 6
   2.3 – Has the Risk Management Framework Worked to Reduce Miner Injuries? 7
   2.4 – The Minerals Industry Safety and Health Centre (MISHC).....................................8
3.0 – Risk Assessment and Analysis Techniques and Tools 9
   3.1 – Risk Assessment Techniques 9
   3.2 – Risk Analysis Techniques and Tools 11
4.0 – Elements of an MHRA 17
   4.1 – Risk Assessment Design (Scoping) 17
   4.2 – Risk Assessment Team 17
   4.3 – Risk Assessment 18
   4.4 – Effectiveness of Controls 20
   4.5 – Audit and Review 22
5.0 - MHRA Pilot Studies at US Underground Mining Operations 23
   5.1 – Rock Reinforcement Process Risk Assessment Case Study 24
   5.3 – Spontaneous Combustion Causing Fire/Explosion Risk Assessment Case Study....33
   5.4 – Underground Workshop Fire Risk Assessment Case Study 40
   5.5 – Water Inundation Risk Assessment Case Study..49
   5.6 – Escapeway Egress Blockage Risk Assessment Case Study 58
   5.7 – Natural Gas Ingress Risk Assessment Case Study 67
   5.8 – Conveyor Belt Fire Risk Assessment Case Study 77
   5.9 – Longwall Gate Entry Track Fire Risk Assessment Case Study 87
   5.10 –Change of Mining Method Risk Assessment Case Study 95
6.0 – Lessons Leaned 110
   6.1 - The Scoping Document 110
   6.2 - The Risk Assessment Team 110
   6.3 – Important Risk Assessment Tools and Techniques 111
   6.4 - The Risk Assessment Team Outputs (Identified Controls)..................................112
   6.5 – Documentation 112
7.0 – Success of Risk Assessment Case Studies 114
   7.1 – Existing Risk Management Culture...114
   7.2 – Risk Assessment Design 115
   7.3 – Risk Assessment Team 115
   7.4 - The Risk Assessment Process 116
   7.5 – The Extent of Existing Controls 116
   7.6 – The Quality of New Ideas 117
8.0 – Future Use of the MHRA Process in Mining 118
9.0 – References 121
APPENDIX B – Action Plan of New Ideas...127

APPENDIX C – Risk Register...128
APPENDIX D - Risk Management Culture and Self-Assessment......................................129

Illustrations

Figure 1 - Cumulative multiple fatalities over the last 10 years in the US Minerals Industry. 4
Figure 2 - Principal risk management framework used in Australia. 7
Figure 3 - The running three-year underground mine fatality rates for Australia and the US. 8
Figure 4 - An example of a WRAC risk ranking form. 11
Figure 5 - The Preliminary Hazard Analysis (PHA) Form. 12
Figure 6 – Item-by-item risk assessment worksheet for FMEA...13
Figure 7 - Process analysis form for a HAZOP..14
Figure 8 - Bow Tie Analysis (BTA) method...15
Figure 9 - Example MHRA team structure (MISHC, 2007). 18
Figure 10 - Drill used to install rock reinforcement. 26
Figure 11 - Photograph of shock cords similar to the ones used at the study site........................31
Figure 12 - Bleederless ventilation system used at the study mine to control spontaneous combustion. 34
Figure 13 - Distribution of prevention controls and recovery measures for the spontaneous combustion causing fire/explosion risk assessments..37
Figure 14 - Map showing the location of the maintenance pit with respect to track haulage, ventilation stoppings, and intake shaft. 40
Figure 15 - Distribution of prevention controls and recovery measures for the underground workshop fire risk assessment...45
Figure 16 - Location of Mines Ea and Eb and adjacent water-filled abandoned mine and water/gas filled adits..49
Figure 17 - Location of geographic boundaries of the risk assessment. Thick lines define the boundaries between the abandoned mines and the current projections for Mines Ea and Eb.......51
Figure 18 - Graphical depiction of the BTA used in the inundation risk assessment. 53
Figure 19 - Techniques used to find the location of water-filled old mine workings. 54
Figure 20 - Distribution of prevention controls and recovery measures for the water inundation risk assessment. 57
Figure 21 - Escapeways, roof falls and recent roof cracks found at the mine.............................58
Figure 22 - Six segments of the mine's escapeway system. 60
Figure 23 - Roof Fall Risk Index (RFRI) measured in the mine's escapeways...........................61
Figure 24 - Distribution of prevention controls and recovery measures for the escapeway egress blockage fire risk assessment. 66
Figure 25 - Mine Ga layout showing the location of active faces, the slope and panels within one active level..67
Figure 26 - The risk assessment team created this schematic to illustrate the hazards and pathways related to the study mines. 69
Figure 27 - Detailed view of the ventilation circuit used at Mine Ga. 70
Figure 28 - BTA for mining into an existing oil/gas well top event. Potential causes are listed on the left side of the bow tie with potential recovery measures on the right...................................71
Figure 29 - Distribution of prevention controls and recovery measures for the natural gas inundation risk assessment. 76

Figure 30 - Mine H layout showing the location of the conveyor belt and working faces. 77
Figure 31 - Segments of the conveyor belt system. 78
Figure 32 - Conditions within the cribbed area of conveyor belt Segment #2. 82
Figure 33 - Distribution of prevention controls and recovery measures for the conveyor belt fire risk assessment. 86
Figure 34 - Site conditions at Mine I showing the 3-entry development panel with direction of air flow. 87
Figure 35 - Distribution of prevention controls and recovery measures for the longwall gate entry track fire risk assessment. 90
Figure 36 - Diagram of Captive Cut-and-Fill mining method. 95
Figure 37 – A flow chart of the basic stope proposal and mine planning process. 105
Figure 38 - Distribution of prevention controls and recovery measures for the captive cut-and-fill change of mining method risk assessment. 109
Figure 39 - Percentage of the total controls by category. 112
Figure 40 - Steps along the path to an improved safety culture. 132

Tables

Table 1 - Hazard types associated with multiple fatality events in the US Minerals Industry, 1997-2007. 5
Table 2 - A generalized risk matrix used in many qualitative risk analysis techniques. 9
Table 3 - Examples of variable scales used to determine the maximum reasonable consequence associated with different kinds of unwanted events. 12
Table 4 - Examples of variable scales used to determine the likelihood of occurrence for different kinds of unwanted events. 12
Table 5 - Examples of variable scales used to define the effects of exposure on risk. 15
Table 6 - A method to determine the total exposure using a 5 x 5 matrix. 16
Table 7 - Estimation of overall likelihood by combining the estimates of likelihood and total exposure. 16
Table 8 - The combinations of maximum reasonable consequence and the likelihood of the maximum reasonable consequence to establish the most likely consequence level. 16
Table 9 - 5 x 5 risk ranking matrix. 16
Table 10 - Categorizing the location and magnitude of the worst hazards using the energy approach. 19
Table 11 - Control categories based on risk reduction effectiveness. 20
Table 12 - Characteristics of the 10 MHRA case study sites. 23
Table 13 - WRAC of the initial steps in the rock reinforcement process. 26
Table 14 - A 5x5 risk matrix used to rank risk in the rock reinforcement process WRAC. 27
Table 15 - New ideas for preventing rock reinforcement selection and installation failures. 27
Table 16 – Potential heat sources and conditions of the atmosphere in the gob needed to cause a fire or an explosion. 35
Table 17 - Existing key prevention controls for the spontaneous combustion risk assessment. ... 36
Table 18 – Existing key recovery measures for the spontaneous combustion risk assessment. ... 37
Table 19 - New ideas for mitigating risk of spontaneous combustion. 38
Table 20 - Fire hazards consisting of potential heat and fuel sources. 41

Table 21 - Preliminary Hazard Assessment (PHA) form for the underground workshop fire risk assessment case study. 42
Table 22 – Existing key prevention controls for the underground workshop fire risk assessment. 43
Table 23 - Existing key recovery measures for the underground workshop fire risk assessment. 46
Table 24 - New ideas for an underground workshop fire risk assessment. 47
Table 25 - Consequences of different inundation mechanisms. 52
Table 26 - Summary of existing prevention controls and recovery measures for a potential mine inundation. 55
Table 27 - New ideas proposed by the risk assessment team for preventing or recovery from an inundation at Mines Ea and Eb. 56
Table 28 – Fire hazards consisting of potential fuel and ignition sources. 60
Table 29 - Risk ranking of potential threats grouped by escapeway segment. 62
Table 30 - A 4 by 5 risk matrix for ranking the potential threats. 62
Table 31 - Existing prevention controls and recovery measures for a loss of emergency escapeway at Mine F. 63
Table 32 - New ideas proposed for preventing or recovery from a loss of emergency escapeway at Mine F. 65
Table 33 – Existing key prevention controls for the natural gas ingress risk assessment (left side of the BTA). 72
Table 34 – Existing key recovery measures for the natural gas ingress risk assessment (right side of the bow tie). 73
Table 35 - New ideas for mitigating risk of natural gas ingress. 74
Table 36 - Characteristics of eight conveyor belt segments. 79
Table 37 - Fuel and heat sources along the conveyor belt. 79
Table 38 – Potential unwanted events for the entire conveyor belt system 80
Table 39 – Three-dimensional risk ranking method used at Mine H. 81
Table 40 – The highest priority risks identified by the WRAC. 81
Table 41 - Summary of existing prevention controls and recovery measures from a potential conveyor belt fire. 83
Table 42 – New ideas proposed by the risk assessment team for preventing or recovery from a conveyor belt fire at Mine H. 84
Table 43 - Fuel and heat sources found within a longwall track entry. 88
Table 44 - Important longwall track entry characteristics considered in the risk assessment. 89
Table 45 - List of acceptable and unacceptable consequences from a longwall track fire. 89
Table 46 - New prevention control and recovery measure ideas for the longwall track fire event organized by category. 93
Table 47 - Hazards associated with captive cut-and-fill mining. 98
Table 48 - Priority listing of potential unwanted events associated with phases in the captive cut-and-fill stoping method at Mine J. 99
Table 49 - Risk Matrix used by cooperating mining company 101
Table 50 - Highest ranked unwanted events associated with captive cut-and-fill stoping. 101
Table 51 - Highest priority risks capable of producing a multiple-fatality event. 102
Table 52 - Priority existing prevention controls and recovery measures for equipment fires in the stope and stope access drift. 102

Table 53 - New prevention control and recovery measure ideas for the equipment fire in the intake drift event. 105
Table 54 - New prevention control ideas for stope design and mine planning. 107
Table 55 - An assessment of the adequacy / success of the ten MHRA case studies. 114
Table 56 - Left side BTA for Mine I. 123
Table 57 - Right side BTA for Mine I. 125

UNIT OF MEASURE ABBREVIATIONS USED IN THIS REPORT

F	degrees Farenheight
ft	feet
ft/s	feet per second
hr(s)	hour(s)
in	inch
lb(s)	pound(s)
min	minute
ppm	parts per million

The Application of Major Hazard Risk Assessment (MHRA) to Eliminate Multiple Fatality Occurrences in the US Minerals Industry

By A. Iannacchione, F. Varley and T. Brady
NIOSH

Abstract

Major Hazard Risk Assessment (MHRA)[1] is used to help prevent major hazards, e.g., fire, explosion, wind-blast, outbursts, spontaneous combustion, roof instability and chemical and hazardous substances, etc., from injuring miners. The structured process associated with MHRA helps to characterize the major hazards and evaluate engineering, management and work process factors that impact how a mine mitigates its highest risk. The National Institute for Occupational Safety and Health (NIOSH) studied the application of this technique to US mining conditions through a field-oriented pilot project. Risk assessment teams used in the pilot project were primarily composed of mining company personnel. Ten case studies were performed over a wide cross-section of mines. These mines were representative of the important mining commodities in the US minerals industry, i.e. coal, metal, non-metal, and aggregate. Also, the sizes of the mines ranged from small to large and were located across the country.

The ten case studies demonstrate that most US mines have the capability to successfully implement an MHRA and that the MHRA methodology produced additional prevention controls and recovery measures to lessen the risk associated with a select population of major mining hazards. The basic ingredient for a successful MHRA is the desire to become more proactive in dealing with the risks associated with events that can cause multiple fatalities. A successful outcome is marked by a thorough examination of existing prevention controls and recovery measures. When pressed to consider more controls to further mitigate the risk, a well-staffed risk assessment team was able to identify additional controls. For these mining operations, it was important to add additional controls, even if they were not required by existing mining regulations, to lower the risks associated with the major hazards under consideration. If a mining operation is not willing to commit its best people to an MHRA or will not provide them with sufficient time to see the process through to its conclusion, the MHRA output may prove to be useless. Additionally, if a mining operation is not prepared to discuss its major hazards in an open and honest fashion and to present the findings of the risk assessment in a written report, the MHRA output will be unclear, and attempts to monitor or audit important controls may not be possible. A MHRA is most effective when the mining operation possesses 1) a proper understanding of its hazards, 2) experience with informal and basic-formal risk assessment techniques, 3) proper facilities, machinery and equipment, 4) suitable systems and procedures that represent industry Best Practice, 5) appropriate organizational support with adequate staff, communications and training, 6) a formal and thorough plan for emergency response, and 7) a

---

[1] Also referred to as Principal or Catastrophic Hazard Risk Assessment.

safety risk management approach that is promoted and supported at all levels of the organization.

Executive Summary

Major Hazard Risk Assessment (MHRA) is a process used to evaluate hazards that can cause great harm to a mining operation and its workers if they are not adequately controlled. NIOSH evaluated the MHRA process at ten mining operations. The general consensus was that the MHRA process provided information considered beneficial for a safer work environment. Three of the ten case studies are rated as performing a more-than-adequate risk assessment, five as adequate, and two as less-than-adequate. The degree of success was influenced by the existing risk management culture at the mining operation, the design of the risk assessment, the performance of the risk assessment team, the character of the risk assessment process, the extent of the existing controls, and the quality of the new ideas. Lessons learned focused on improving the scoping document, the need to adequately train the risk assessment team, the important risk assessment tools and techniques, methods to assess the quality and character of the risk assessment team outputs, and the significance of the documentation process.

Fundamental to successful utilization of risk assessment in the MHRA process is company support to form a team with the capability and intentions to address all hazards. It is critical that the risk assessment be designed to capture the strengths of the MHRA approach in order for it to be successful. The strengths of the MHRA approach are its ability to
1. set clear direction to solve specific high-risk problems,
2. focus on priority concerns,
3. establish involvement and commitment from a wide cross-section of the mine's work force,
4. decrease potential losses for mining operations,
5. help to build teams to solve major mining issues,
6. go beyond merely complying with existing mining standards and regulations, and
7. focus upper management attention on issues existing at the operational level.

Conversely, the MHRA approach is unlikely to prove successful if the following issues or concerns take precedence during a risk assessment:
1. inappropriate focus on changes within the existing way the mine conducts business,
2. time taken away from activities directly related to production,
3. focus on additional time constraints being placed on a mining operation's "best people,"
4. the cost of implementing new prevention controls and recovery measures,
5. inappropriate alteration of a mining operation's priorities,
6. need for there to be an existing risk management structure to build upon, and
7. need for an openness in management / labor communications.

This NIOSH pilot project demonstrated that US mines have the capability to successfully implement an MHRA and that the basic requirement for a successful MHRA is the desire to become more proactive in reducing risks associated with events that can cause multiple fatalities. An MHRA can be most effective when the mining operation possesses a proper understanding of its hazards, has some experience with risk assessment techniques, uses systems and procedures that represent industry best, or attains wide organizational support for the MHRA activity.

The power in the MHRA process comes from the risk assessment team as it examines new ideas that will help to further reduce risk. These new ideas are presented to management in the form of an Action Plan. This Action Plan is contained within a written document that summarizes the risk assessment team's actions and is presented to management. The Action Plan also suggests that management assign a responsible person to evaluate each of these new potential controls and recovery measures in a more in-depth manner. Management can then select the new ideas most appropriate for their mine.

Acknowledgement

The authors would like to thank each of the companies that participated in the pilot project. It is not easy for companies to allow persons outside their organization to examine hazards within their mining operations. The companies involved did this without reservation and for that we are grateful. We would also like to thank NIOSH management for their support and encouragement. Lastly, we wish to recognize the guidance and efforts of Professor Jim Joy[2] in facilitating the MRHAs. He was forever teaching and all were his interested students.

1.0 - Introduction

The reoccurrence of multiple fatality events in the US Minerals Industry supports the need for improvements in the way major hazards are identified, assessed and managed. Many solutions to reduce mining disasters have been proposed including additional regulations, improved training, more reliable equipment, and better technology. In December of 2006, the National Mining Association's Mine Safety Technology and Training Commission stated that a new paradigm for ensuring safety in underground mines was needed. The Commission recommended that the industry consider a systematic and comprehensive risk management approach (Grayson et al., 2006). In March of 2007 during a congressional hearing, the NMA announced its support of a risk assessment based approach for the mining industry (Watzman, 2007). In another congressional hearing, Davitt McAteer asked that Best Practices be prepared which could be used to hold mine operators to a higher standard of care, i.e. risk assessment and risk control (McAteer, 2007).

The elimination of multiple fatality events is arguably one of the most important safety issues facing the US Minerals Industry. Ten case studies are presented that use a range of practices to lower the risk from site-specific major hazards. These practices ranged from standard to those that are leading the industry. This paper evaluates how the use of Major Hazard Risk Assessment (MHRA) might help to eliminate multiple fatality events. The MHRA process was developed by the Australian mining industry over the last decade as a means of mitigating catastrophic hazards from its mining operations.

Most case studies were viewed as successful by the quality of the barriers, controls and recovery measures produced during the risk assessment and the responses of the individual risk

---

[2] Minerals Industry Safety and Health Centre, University of Queensland, Australia

assessment teams. However, some unsuccessful outcomes were also observed. Of particular importance was the need to communicate risk management principles to both management and labor, the knowledge of the hazards possessed by the team used to conduct the risk assessment, and the ability of the existing and proposed prevention controls and recovery measures to go beyond simply complying with existing government standards and regulations. This report should be viewed as a guidance document to provide the industry with information and tools needed to implement a successful MHRA program.

1.1 – Trends in Managing Major Mining Hazards

A proven way to manage the many hazards associated with mining is to characterize the risk they present and put into place controls that will lower these risks to acceptable levels. Typically risk acceptability is characterized by managing risk to as-low-as is reasonably achievable (ALARA) or as-low-as is reasonably practicable (ALARP). An industry's ability to manage risk is often measured by its injury and illness rates. If rates are falling and lower than those of other developed countries or comparable to other similar high-hazard industries, then that industry is considered to have demonstrated a proficiency in managing hazards. One way to examine the US Minerals Industry's proficiency in managing its risks is to examine multiple fatality trends. Over the last 10 years, there have been 18 multiple fatality events in the US, fatally injuring 67 miners (*Figure 1*).

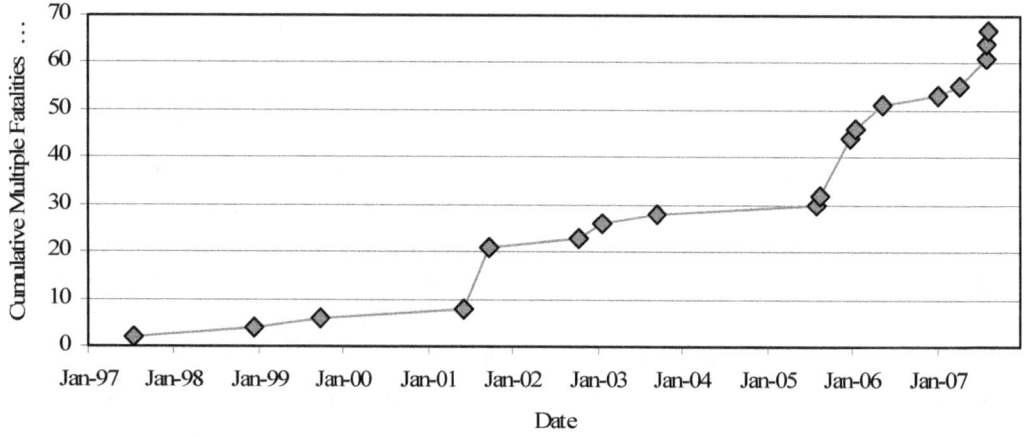

Figure 1 - Cumulative multiple fatalities over the last 10 years in the US Minerals Industry.

Sixteen of the 18 multiple fatality events occurred in coal mines and 15 at underground operations. The most frequent event type is strata instabilities with 8 events fatally injuring 21 miners (*Table 1*). Explosions were involved in fewer events, 4, but had the highest number of fatalities, 33. Powered haulage, fire, heat strain, equipment failure and slip or falls of persons are other examples of major hazards in the US Minerals Industry. These data suggest that major hazards exist in our nation's mines capable of causing multiple fatality events. They also support the need for additional actions to lesson the impact of these hazards on miner safety.

Table 1 - Hazard types associated with multiple fatality events in the US Minerals Industry, 1997-2007.

Hazard Type	Events	Fatalities
Strata Instabilities	8	21
Explosions	4	33
Powered Haulage	2	4
Fire	1	2
Equipment Failure	1	2
Heat Strain	1	2
Slip or Fall of Person	1	3

2.0 – Minerals Industry Risk Management

Risk management systems have been used in many industries to manage inherent hazards in their business. In fact, some countries mandate risk management approaches in their minerals industries. Others, like the US, produce technically detailed regulations, often reacting to a particular disaster, with the purpose of prescribing specific industry actions. By evaluating these different approaches to risk management, an assessment can be made of the impact of the risk management framework on miner injuries.

2.1 – The US Experience

Prescriptive mining standards rely on existing normalized rules, largely based on past-experience and current Best Practices, to mandate safety standards. These standards can produce lengthy and detailed regulations. Changing technology and mining conditions require the regulations to be constantly reviewed and, on occasion, modified. However, prescriptive standards are sometimes incapable of dealing with hazards associated with specialized and dynamic mining conditions. They can also produce a culture of compliance that does not necessarily emphasize leading practice. The above process could potentially lead to a reactive approach towards hazards.

Alternatively, regulatory standards with a General Duty Clause[3] require employers, suppliers and employees to provide, design for and adhere to reasonable activities ensuring that workers are protected. Many industries, e.g. nuclear, petrochemical, environmental, have used structured risk management approaches to develop proactive approaches in managing their risks. In these industries, safety plans often focus on a local site's approach toward assessing risks and mitigating these risks through targeted controls.

The experience of these industries provides an opportunity to examine how a risk management approach may help to eliminate major hazards in US mineral industries. In addition, many of the

---

[3] In commonwealth countries including the UK, Canada, Australia, and New Zealand, a similar clause is often found in regulation referred to as the Duty-of-Care.

commonwealth countries, e.g. United Kingdom, Canada, Australia, New Zealand, have used risk management as their guiding principles for current mining standards. Perhaps the country that has had the most extensive transition, from a prescriptive-based health and safety philosophy to a more proactive, duty-of-care philosophy, is Australia.

2.2 – The Australian Experience

In Australia, the minerals industry began its movement towards risk-based management systems in the mid-1990s, shortly after the Moura coal mine explosion fatally injured 11 miners (Hopkins, 2000). Later in 1996, the Gretley coal mine inundation reinforced the drive for change. As a result, industry began using risk analysis methods to mitigate certain key hazards, e.g. fires, explosions, inundations, spontaneous combustions, etc. Later, the various regulatory bodies in Australia began to mandate safety management plans for principal hazards. In New South Wales, the Chief Inspector of Coal Mines (NSWDPI, 1997) published a risk management handbook that offers a process to anticipate and prevent circumstances which may result in occupational injury or death. Queensland followed (QDME, 1998 and QMC, 1999) with its own standard. In Western Australia, where the largest concentration of metal mining occurs, duty-of-care legislation was enacted in 1994; however, risk management approaches saw less application until recently (CMEWA, 2003). Most of these regulations require mines to perform some form of risk assessment on a regular basis to address the possibility of unwanted events such as spontaneous combustion, gas outbursts, explosions, air blasts, inundations and roof falls. In addition, mine managers are generally expected to demonstrate competency in risk-based management systems through training and certification. As is evident from this discussion, the Australian Minerals Industry performs MHRA, in part, because it is mandated.

In response to these different approaches to duty-of-care regulations, an Australian Standard on Risk Management (Standards Australia, 2004) was established, providing an important risk management framework (*Figure 2*). First, hazards are identified by the location, nature and magnitude of energies present within a mine. The risks these hazards present are then identified and assessed. Next, the mine operator decides whether to eliminate, mitigate or tolerate these mining hazards. Typically it is most effective to eliminate hazards early in the life of a mine when design activities are the most prevalent. Mitigation actions can consist of equipment, materials, rules, methods, competencies, labels or other mechanisms to control hazards. If a hazard is tolerated then administrative controls, specialized training or recover measures are used to minimize losses. As these actions are taken, their performance must be monitored. This is typically done on a regularly scheduled basis and changes are made to the process as needed.

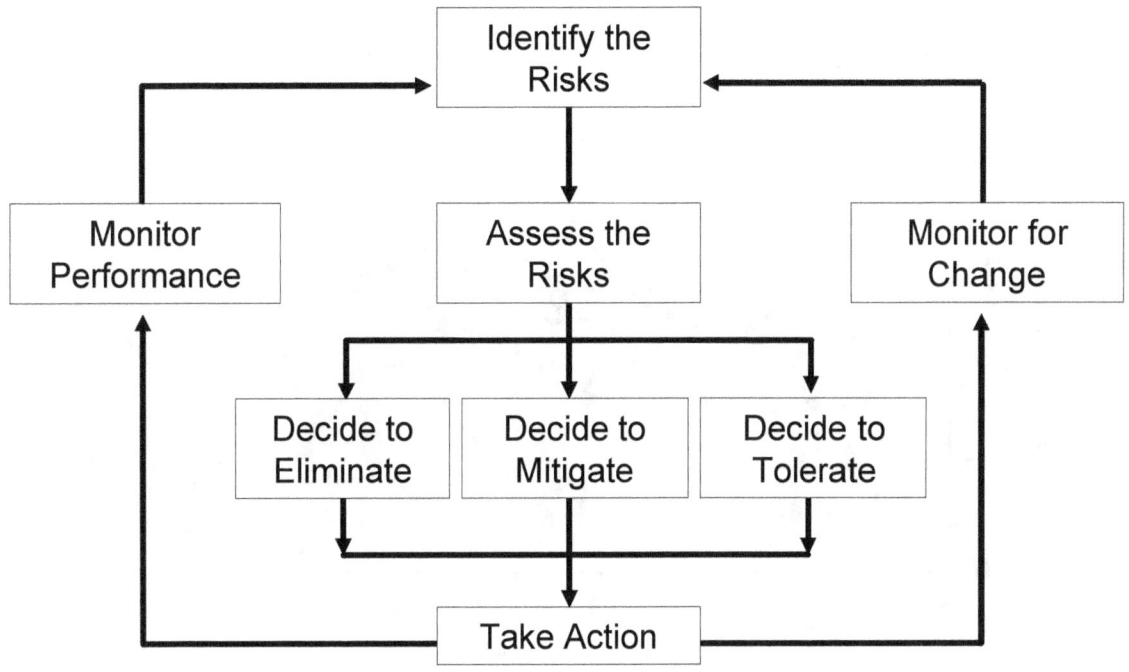

Figure 2 - Principal risk management framework used in Australia (Standards Australia, 2004).

2.3 – Has the Risk Management Framework Worked to Reduce Miner Injuries?

In Australian underground coal mining, following an increasing fatality rate in the early 90s, there has been a marked decrease in these same rates over the last 10 years with a dramatic drop in the last three years (
Figure 3). Unfortunately the fatality rate for US underground coal mining has not seen this same drop. In underground Australian metalliferous mining, a significant decrease only occurred in the last 5 years. Also, the US metal/non-metal underground mining fatality rate, once well above the level for US underground coal, is now consistently lower than coal. It is also worth noting that there has not been a coal mine multiple fatality event in Australia since the Gretley inundation in 1996; however, there have been two multiple fatality events in metalliferous mines (Bronzewing and North Parks) during this same time frame. The general downward trends in Australian underground mining fatality rates are, in part, attributed to the introduction and acceptance of risk-based management systems. It should be noted that the Australian experience also demonstrates the need for continual improvement and government oversight in response to improperly managed risk-based management systems (Freeman, 2007).

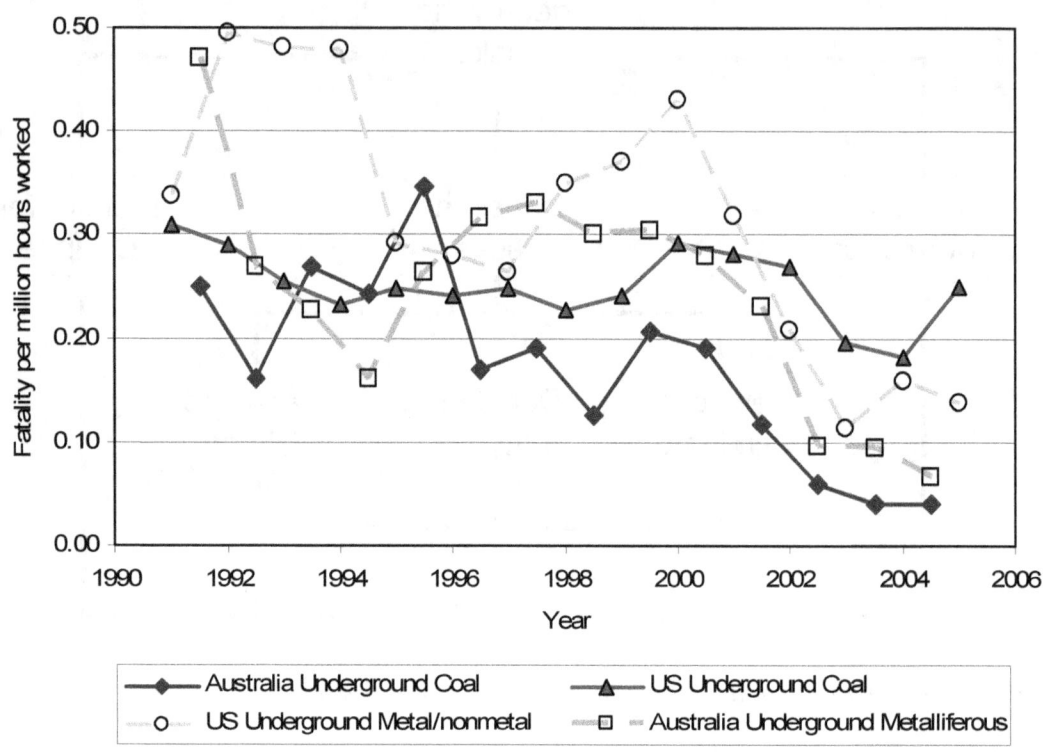

Figure 3 - The running three-year underground mine fatality rates for Australia and the US.

2.4 – The Minerals Industry Safety and Health Centre (MISHC)

To assist in evaluating the MHRA approach, NIOSH sought help from a leading Australian institution involved in implementing mineral industry risk management programs. Several major companies and the government of Queensland help to form the Minerals Industry Safety & Health Centre (MISHC) at the University of Queensland in 1998. This centre conducts research and education as well as develops industry resources on risk management topics. MISHC has developed a Minerals Industry Risk Management (MIRM) model for achieving "safe production" (Joy, 2006), requiring managers to be knowledgeable about hazards inherent in their operations and to follow a logical framework to define effective barriers or controls. The MHRA approach used in the ten NIOSH case studies follows the approach taught by MISHC and is indicative of the approach used by most Australian mining companies to manage hazards with multiple fatality potential. MISHC is also responsible for maintaining the Minerals Industry Risk Management Gateway or MIRMGATE (www.mirmgate.com). This site was found to be a good source for Best Practice hazard management guidelines, lessons learned and innovations.

3.0 – Risk Assessment and Analysis Techniques and Tools

Risks are determined in terms of the likelihood that an uncontrolled event will occur and the consequences of that event occurring.

Risk = Likelihood of occurrence × consequence

The above relationship is used in both qualitative and quantitative risk analysis methods. A quantitative risk analysis method is a probabilistic estimation of risk where risk is calculated as a continuous series from high to low. A qualitative risk analysis method is a basic estimation where risks are typically ranked from high to low. Qualitative methods rely on a risk matrix similar to that demonstrated in *Table 2* where qualitative categories are defined, i.e. low-to-high, unlikely-to-likely, etc.

Table 2 - A generalized risk matrix used in many qualitative risk analysis techniques.

		Likelihood of Occurrence		
		High value	Medium value	Low value
Consequence	High value	High risk	Moderate risk	
	Medium value			
	Low value			Low risk

Risk assessment and analysis techniques and tools consist of a systematic, logical set of actions used to identify hazards, assess risk, and implement controls to mitigate high-risk conditions. These techniques and tools can be described by their levels of formality, the types of analysis performed, and the work processes they are attempting to address.

3.1 – Risk Assessment Techniques

The most fundamental risk assessment activity, called an informal risk assessment, occurs when workers are asked to think about the hazards in the workplace before work commences, determine what could go wrong, and report or fix the hazards. More formal risk management activities require structured procedures, often focusing on work processes that involve multiple levels of an organization. These activities are practiced at some mines and are typically organized by an operations safety official and developed with the help of individuals familiar with the work practice in question. Higher level risk management activities focus on major mining hazards or on major changes in the mining operations involving the entire organization, such as reopening a mine, moving to a new location within the mine, and utilizing a new mining technique or process.

3.1.1 – Informal Risk Assessment Techniques

Most informal risk assessment techniques consist of multiple steps where the worker is asked to look for hazards, determine the significance of the hazard, and take some action to mitigate the risk. Many systems have been proposed and are widely used in mining. Examples include, but are not limited to:

- Stop-Look-Analyze-Manage (SLAM) asks workers to stop and consider the work process before it is started, examine the work environment, analyze the work process, and manage the risk,
- Take-Two for Safety calls for persons to take 2 minutes to think through a job before it starts,
- Five-Point Safety System compels employees to take responsibility for the safety within workplace,
- Take Time, Take Charge requires miners to stop, think, assess and respond to hazards in their workplace.

3.1.2 – Basic-formal risk assessment techniques

Basic-formal risk assessment techniques are characterized by the requirement to follow a structured process that occurs prior to performing specific higher risk work activities. These techniques also require documentation that allows management to monitor and audit individual risk assessment activities. The most commonly used basic-formal risk assessment technique is the Job Safety Analysis (JSA). A JSA typically leads to development of Standard Operating Proceducres (SOP) that define how to best approach a task considering the hazards identified in the JSA.

A JSA is a technique used to identify, analyze and record the specific steps involved in performing a work activity that could have hazards associated with it. JSAs are typically performed on work processes with the highest risk for a workplace injury or illness. It is essential that all actual or potential safety and health hazards associated with each task are identified and that actions or procedures for performing each step that will eliminate or reduce the hazard are documented and recorded. Other techniques similar to JSAs include Job Hazard Analysis (JHA), Critical Task Analysis (CTA), and Job Hazard Breakdown (JHB).

An SOP is a set of instructions that act as a directive, covering those features of operations that lend themselves to a standardized procedure. An SOP is typically a set of instructions or steps a worker follows to complete a job safely and in a way that maximizes operational and production requirements. SOPs can be written for work processes by the individual or group performing the activity, by someone with expertise in the work process, or by the person who supervises the work process.

3.1.3 – Advanced-formal risk assessment techniques

Advanced-formal risk assessment techniques require the use of a structured approach that incorporates one or more risk analysis tools (see Section 3.2) and produces a documented assessment of the risk associated with unwanted events. MHRA, the subject of this investigation, is an advanced-formal risk assessment technique. An MHRA can focus on a single major hazard, all the relevant major hazards, or an important change of mining method at a mining site. One study demonstrates the complexity that a change of mining method can bring to the risk assessment. In this case, a full week of effort from a large team was needed with multiple risk analysis tools. All other MHRAs studied are focused on a single hazard and were completed in 1 to 3 days.

3.2 – Risk Analysis Techniques and Tools

When conducting an MHRA several risk analysis techniques and tools may be needed. A brief description of the most common tools follows.

3.2.1 – Workplace Risk Assessment and Control (WRAC)

The Workplace Risk Assessment and Control (WRAC) tool is a broad-brush risk ranking approach, allowing the user to focus on the highest risk. As applied to a MHRA, this structured preliminary analysis begins by breaking down the mining process associated with the potential major hazards at the mine in some logical manner. This is often accomplished using a flow chart or process mapping technique where the potential major hazards of each step in a work process are identified. The mining process could be a breakdown of a major project or a geographical breakdown of the underground mine. JSAs and SOPs can be used as a framework for the WRAC analysis.

After preliminary analysis, the team then considers each breakdown segment of the mining process and identifies the potential unwanted events associated with the identified hazards (*Figure 4*). The likelihood and consequence of each stage are determined using some variation of a risk matrix, followed by a risk rating calculation.

Part of mine, phase of mining, etc.	Potential unwanted event	Consequence	Likelihood	Risk rating
	↕			

Figure 4 - An example of a WRAC risk ranking form.

Prior to ranking the hazard, the team must come to an agreement on how to categorize the consequences for consistency purposes. Consequences should be considered as either the maximum likely or the maximum potential consequence. For example, while the maximum potential consequence of a slip/fall is a fatality, the maximum likely consequence is a severe injury. Variable scales are often used when determining the maximum reasonable consequence associated with different kinds of unwanted events. *Table 3* provides some examples of the maximum reasonable consequence for safety, equipment, production and environmental risks. This table also provides a scale for determining the maximum reasonable consequence of a specialized safety event, in this case a mine fire.

Table 3 - Examples of variable scales used to determine the maximum reasonable consequence associated with different kinds of unwanted events.

	Safety	Equipment	Production	Environment	Mine Fire
1	Multiple fatalities	> $ 5 M	1 Week	> $ 5 M	Huge fire
2	1 Fatality	$ 1 M	1 Day	$ 1 M	Major fire
3	Major lost-time injury (LTI)	$200 K	1 Shift	$200 K	Moderate fire
4	Avg LTI (4-5 days)	$50 K	1 Hour	$50 K	Small fire
5	Minor injury (1 day or less)	< $ 10 K	<1 Minute	< $ 10 K	Smoldering

LTI = lost time injuries
M = million
K = thousand

The ranking of likelihood will be influenced by the choice of consequences. There is no correct choice, but there is a need to be consistent in the application of ranking across the exercise. Examples of variable scales used to determine the likelihood of different kinds of unwanted events is given in *Table 4*.

Table 4 - Examples of variable scales used to determine the likelihood of occurrence for different kinds of unwanted events.

	Based on Maximum Reasonable Consequence			Based on the Events / Year
1	Common	Highly likely	Expected	> 10
2	It has happened	Likely	High	1 to 10
3	Possible	Possible	Moderate	0.1 to 1
4	Unlikely	Unlikely	Low	0.01 to 0.1
5	Almost impossible	Very unlikely	Not Likely	< 0.01

3.2.2 – Preliminary Hazard Analysis

The Preliminary Hazard Analysis (PHA) is another broad-brush risk ranking approach. Like the WRAC, this tool identies all potential hazards and unwanted events that may lead to miner injuries and ranks the identified events according to their severity. Its main purpose is to identify those unwanted events that should be subjected to further, more detailed risk analysis. Once the potential unwanted events are risk ranked by the team, they can be prioritized so that the highest risk unwanted event is listed first and so on. The technique or form for the PHA method is shown in *Figure 5*.

#	Description of potential unwanted event	Total Exposure	Likelihood	Most Likely Consequence	Risk Rank
1					
2					
3					
4					
Etc.					

Figure 5 - The Preliminary Hazard Analysis (PHA) Form.

3.2.3 – Failure Modes, Effects and Analysis (FMEA), also FMECA

Generally, an FMEA is used to determine where failures can occur within hardware and process systems and to assess the impact of such failures. For each item, the failure modes of individual items are determined, effects on other items and systems are recognized, criticality is ranked, and the control is identified (*Figure 6*).

Item	Failure Mode	Effects on		Likelihood (L)	Consequence (C)	Criticality (LxC)	Control
		Other items	System				
				↕			

Figure 6 – Item-by-item risk assessment worksheet for FMEA.

Robertson and Shaw (2003) provide an example of the application of the FMEA approach where the risks to the environment, workers and the public associated with the closure of a mine were identified. This was accomplished by developing a FMEA worksheet for potential unwanted events post-closure of the mine.

3.2.4 – Fault / Logic Tree Analysis (FTA/LTA) and Event / Decision Tree Analysis (ETA/DTA)

The Fault and Logic Tree Analysis are systematic, logical developments of many contributing factors to one unwanted event. The FTA evaluates the one unwanted event while the LTA evaluates a wanted outcome. With both tools it is necessary to first clearly define the top event, followed by an analysis of the major potential contributing factors. Each contributing factor is broken down into discrete parts. A logic tree can be used to test the analysis with the use of "and/or" gates. Factors can be ranked from major to lesser. The product of the analysis is a deductive list of potential hazards. This tool is well-suited to quantitative risk analysis techniques when probabilities for each factor can be assigned.

Systems engineering and operations research approaches use a decision tree (or tree diagram) to help examine the decision. Event and decision tree analysis (ETA or DTA) uses graphical models to examine the consequence of decisions. A decision tree is used to identify the strategy most likely to produce a desired outcome. In the tree structures, leaves represent classifications and branches represent conjunctions of features that lead to those classifications. These tools are appropriate for establishing lines of assurance and determining their success and failure in preventing accidents.

3.2.5 – Hazard and Operability Studies (HAZOP)

Hazard and Operability Studies or HAZOPs have been used extensively in the chemical industries to examine what impact deviations can have on a process. The basic assumption when performing a HAZOP is that normal and standard conditions are safe and hazards occur only when there is a deviation from normal conditions. A HAZOP can be conducted during any stage of a project although it is most beneficial during the later stages of design. Typically a process or instrumentation diagram is used to trace the properties of materials or products through a plant by breaking down the process node by node (*Figure 7*). The properties can be flow, level, pressure, concentration or temperature. What-if guidewords are used to identify possible deviations. A HAZOP typically lacks a risk calculation.

Process Unit:				
Node:		Process Parameter:		
Guide	Deviation	Consequence	Causes	Suggested Action
		↕		

Figure 7 - Process analysis form for a HAZOP.

3.2.6 – Bow Tie Analysis

The Bow Tie Analysis (BTA) was developed by Shell Oil in the 1980s as part of its Tripod package of concepts and tools for managing occupational health and safety in its business. The "Top Event" in the BTA is a statement about the initiating event that might lead to the major consequence (*Figure 8*). Threats (also referred to as potential causes) are discussed and controls examined that could mitigate the hazard (left side of the bow tie). Next, the consequences (also referred to as the potential outcomes) of the initiating unwanted event are identified and recovery control measures examined to reduce or minimize the loss (right side of the bow tie).

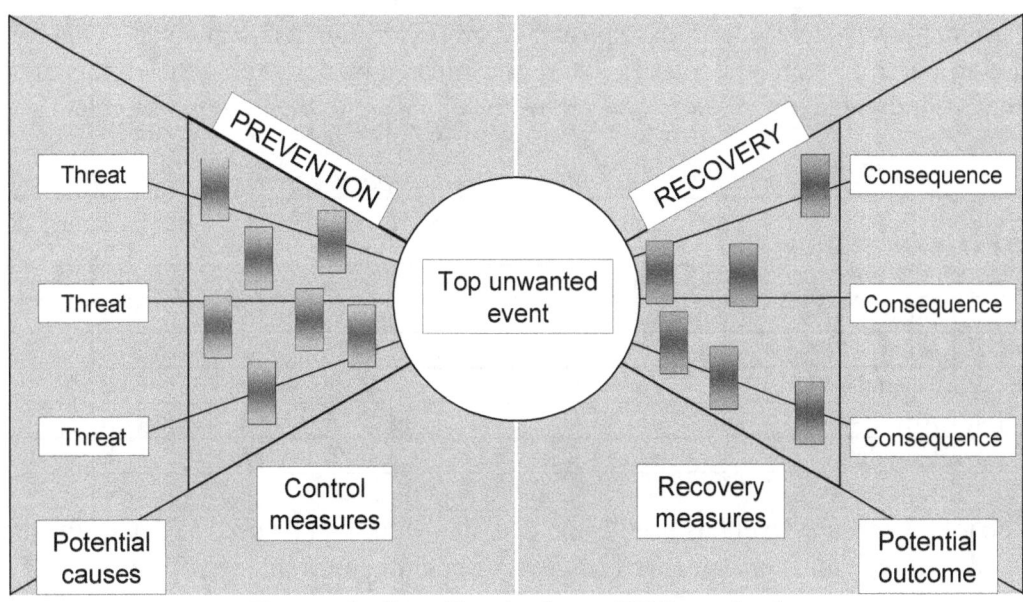

Figure 8 - Bow Tie Analysis (BTA) method.

Together, the prevention controls and recovery measures identified represent a comprehensive list of actions required to adequately control the hazard. Often these actions are assigned to individuals and controlled by the management planning and monitoring tool known as a risk register. A risk register provides for continuity in the way an organization deals with risks even as changes occur in management.

3.2.7 – Work Process Flow Chart

All mining processes have supplies, inputs, processes and outputs. Mining processes are sets of activities that produce a desired outcome. Many of these activities can be thought of as loops. If the outputs are wrong then adjustments are made to the inputs or the process. Defining these work processes in a step-by-step manner produces a flow chart that can be used in risk management. Flow charts are meant to describe a large, sometimes complex, process as small elements. Hazards are easier to identify and characterize with this type of systemic approach.

3.2.8 – Exposure and Risk

When miners are exposed to variable contact with hazards, it is often useful to determine the influence of exposure associated with different work processes or at particular work sites. An example of the variable scales used to define the effects of exposure on risk is given in *Table 5*.

Table 5 - Examples of variable scales used to define the effects of exposure on risk.

	Exposure (% of workforce)	Frequency of exposure
1	Most > 50%	Continuous
2	Many – 30%	Several times/day
3	Several – 10%	Once a day
4	A few – 5%	Weekly
5	Very few < 1%	Monthly

There are many ways to account for exposure when performing a risk analysis. One example is provided in *Table 6*. Here, the total exposure is estimated by combining the effects of the frequency of individual miner exposure versus the exposure to the total workforce.

Table 6 - A method to determine the total exposure using a 5 x 5 matrix.

TOTAL EXPOSURE			Frequency of exposure (*Table 5*)				
			1	2	3	4	5
			Continuous	Several x/day	1/day	1/week	1/month
Exposure (*Table 5*)	1	Most > 50%	A	A	B	C	D
	2	Many – 30%	A	B	C	D	E
	3	Several – 10%	A	C	D	E	E
	4	A few – 5%	B	D	E	E	E
	5	Very few < 1%	C	D	E	E	E

Once the total exposure level has been estimated, this value can be used to determine the overall likelihood (*Table 7*) and consequence (*Table 8*) of potential unwanted events occurring.

Table 7 - Estimation of overall likelihood by combining the estimates of likelihood and total exposure.

LIKELIHOOD OF AN EVENT		Total exposure (*Table 6*)				
		A	B	C	D	E
Likelihood (*Table 5*)	Common	A	A	A	B	C
	Has happened	A	B	B	C	D
	Possible	B	C	C	D	E
	Unlikely	C	D	D	E	E
	Very unlikely	D	E	E	E	E

Table 8 - The combinations of maximum reasonable consequence and the likelihood of the maximum reasonable consequence to establish the most likely consequence level.

MOST LIKELY CONSEQUENCE		Likelihood of the Consequence (*Table 4*)				
		Highly likely	Likely	Possible	Unlikely	Very unlikely
Maximum Reasonable Consequence (*Table 3*)	Multi-fatality	A	A	B	C	D
	1 fatality	A	A	B	C	D
	Serious LTI	B	B	C	D	E
	Avg LTI	C	C	D	E	E
	Minor LTI	D	D	E	E	E

The total probability and consequence of the potential unwanted event are then determined using a 5 x 5 risk ranking matrix (*Table 9*).

Table 9 - 5 x 5 risk ranking matrix.

RISK RANK		Overall Likelihood (*Table 7*)				
		A	B	C	D	E
Overall Consequence (*Table 8*)	A	1	2	3	7	11
	B	3	5	8	12	16
	C	6	9	13	17	20
	D	10	14	18	21	23
	E	15	19	22	24	25

4.0 – Elements of an MHRA

The elements of the MHRA approach used in the case studies are established and published by MISHC (www.mishc.uq.edu.au). During these studies, MISHC personnel assisted NIOSH in conducting the MHRA. Training on the MHRA process was given to participants when possible. A risk assessment scope and the potential team participants were most often identified during these training sessions.

4.1 – Risk Assessment Design (Scoping)

The risk assessment design or scope is best defined prior to the MHRA exercise. Major hazards to be discussed, decisions on risk assessment team membership, and time allotment for the activity are best addressed with a scoping document. This document provides an opportunity to break down the MHRA process into reviewable parts containing the following information:

1. An objective statement that identifies potential major hazards of interest to the mining operation,
2. The boundaries of the mining system or work process,
3. The risk analysis methods and risk assessment tools,
4. The names of potential team members,
5. The time and dates of the MHRA,
6. The location of the MHRA,
7. Determination of the potential data requirements, i.e. in-house safety statistics, MSHA data related to the hazard(s), and similar assessments from the MIRMgate website (www.mirmgate.com),
8. The use of experts from outside the mining operation, and
9. The types of documents that will be produced.

4.2 – Risk Assessment Team

A fundamental part of an MHRA is the risk assessment team. This team must include an appropriate cross-section of knowledgeable persons familiar with the hazards to be investigated (*Figure 9*). The team must be capable of identifying all relevant hazards, unwanted events and possible controls. The process leader is the facilitator who has the appropriate qualifications, knowledge and experience. The facilitator is responsible for following a quality risk assessment process designed to meet the risk assessment scope and is responsible for making sure the team and the process remain focused on a quality output. It is important for the facilitator to act as a teacher and coach without dominating the discussion while making sure to avoid conflict and imbalance in involvement of team members. Open ended questions are often used to elicit participation from the group.

It is also important to consider non-management/labor entities for team participation. Miners responsible for performing tasks that are part of the work processes under review can validate information and provide insight, perspective and ideas that are invaluable to a quality output. These team members are also helpful in communicating adherence to existing prevention controls and recovery measures and in embracing changes brought about by the application of new ideas.

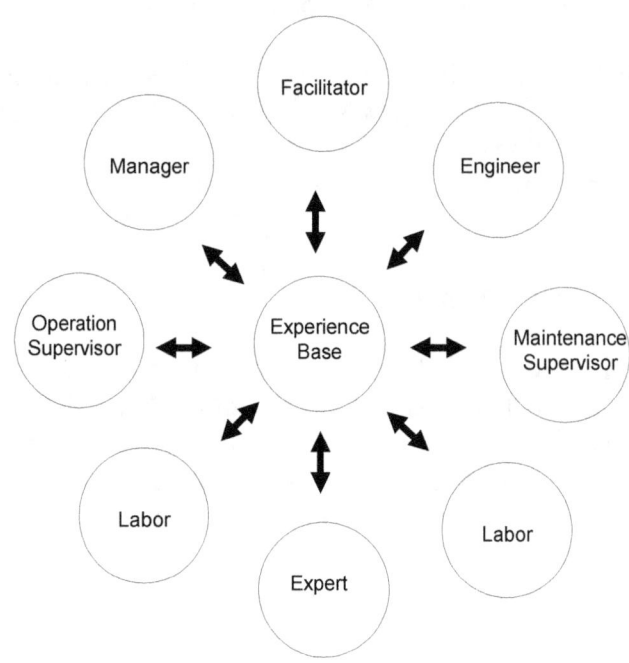

Figure 9 - Example MHRA team structure (MISHC, 2007).

The individuals assigned to the MHRA should fully dedicate their time to this effort during the assessment. It is very important for team members to receive instruction on the MHRA approach prior to the risk assessment. The time allotted for the MHRA should be determined by the complexity of the topic. A focused topic could be done in one day, while more complex topics or a site-wide MHRA could take 3 to 5 days. The venue for the activity is also important. The location should be quiet, free from disruptions, with tables set in a U-shaped pattern to promote discussions and equality among members.

4.3 – Risk Assessment

Five basic steps make up the MHRA process:
1. Identify and characterize major potential mining hazards,
2. Rank potential unwanted events,
3. Determine important existing prevention controls and recovery measures,
4. Identify new prevention controls and recovery measures, and
5. Discuss implementation, monitoring and auditing issues.

4.3.1 – Identify and Characterize Major Potential Mining Hazards

The first step is to identify all relevant hazards or possible problems that could lead to a potential multiple fatality event. If the list is incomplete, the risk assessment will be inadequate. The types of hazard that should be identified are best thought of as uncontrolled releases of energy that have the potential to cause significant harm (Standards Australia, 2004). The energy approach is used to think about what could go wrong specifically at a mining operation. If an accident can be thought of as an uncontrolled release of energy, then it follows that the risk of an accident is higher when the energy is large. An exact measure of energy is not needed, only recognition that it can do serious harm. Potential energies inherent in many mines are contained within the roof and rib of the underground mine, the highwall of the surface mine, the chemical energy in toxic or explosive gases, and the fluid energy of water above or adjacent to the mine workings. The process of mining also brings large-scale energy into the mining environment, i.e. the mechanical energy in mobile and processing equipment, the shocking energy in electrical equipment, the air and hydraulic pressure in fluid systems, etc. Energy that is not completely controlled leads to some level of risk, depending on the likelihood of release and the consequences should the energy be released. When the unwanted release occurs, it can cause serious injuries. *Table 10* is used to list typical sources of energy and to characterize their possible locations and magnitudes.

Table 10 - Categorizing the location and magnitude of the worst hazards using the energy approach.

Hazard (Energy approach)	Location	Magnitude (worst case)

4.3.2 – Rank Potential Unwanted Events

After a comprehensive list of hazards is identified and characterized, a broad-brush risk assessment tool such as the WRAC or PHA is used to risk rank the potential unwanted events. Depending on the topic, the individual hazards should be broken down using a process mapping technique or by the geographic location within the mine. For each step in the work process or for each geographic location within the mine, a likelihood of occurrence and a consequence for each potential hazard are determined. It should be noted that in some MHRA case studies, likelihood is ignored because the consequences of the unwanted event are deemed significant. For all field studies, a qualitative risk ranking procedure is used, integrating some variation of the risk matrix shown in *Table 2*. At the conclusion of this step, the team has successfully ranked the risk. The highest rank risks are almost always unacceptable and the lowest rank risks are often acceptable.

4.3.3 – Determine Important Existing Prevention Controls and Recovery Measures

Additional risk assessment tools are used to help determine what prevention controls and recovery measures are currently being used. In most cases, the BTA or the work process flow chart are excellent tools to conduct detailed analysis of the highest ranked risks. At the end of this step, a detailed list of all existing prevention controls and recovery measures for the hazard in question are documented so they can be monitored and audited on some regular basis.

4.3.4 - Identify New Prevention Controls and Recovery Measures

The same process that identifies existing prevention controls and recovery measures is used to identify new prevention controls and recovery measures. This is a crucial step since it potentially produces a list of actions to be investigated that are capable of further reduction of risks at an underground mine site. It is important for management to consider the merits of each new idea suggested by the risk assessment team. Typically, these new ideas are presented in the form of an action plan.

4.3.5 - Discuss Implementation, Monitoring and Auditing issues

A document is produced at the conclusion of the MHRA that focuses on a description of the hazards examined, the ranks of the potential risks, and a summary of both existing and new prevention controls and recovery measures. The document does not rank the new ideas nor should it attempt to define a specific course of action or recommend a specific design solution for management. A post-risk assessment presentation by the team to management is made to gain acceptance and understanding of the MHRA outcomes. Two key responses from management are needed. First, management needs to make sure that all existing prevention controls and recovery measures identified by the risk assessment team are monitored, audited, or investigated to ensure that unacceptable risks are controlled. Second, all suggested new ideas should be, at a minimum, investigated.

4.4 – Effectiveness of Controls

The important output of the risk assessment team is the list of existing and new controls. Assessing the quality of this output can only be accomplished when the effectiveness of these controls is understood. In this study, the controls were categorized using a hierarchy framework (*Table 11*) used by MISHC personnel. When a hazard is eliminated, the risks associated with the hazard are also eliminated. This should always be the first action of the risk assessment team – to investigate how to eliminate the hazard. However, this is usually difficult to do, since a hazard can owe its origin to many different factors. Some of these factors are poorly understood, while others may represent a condition of business that is perceived to be difficult to change.

Table 11 - Control categories based on risk reduction effectiveness.

Control Category Based on Hierarchy Framework	Major Control Issues	Potential for Human Error	Risk Reduction Effectiveness
Eliminate Hazard (EH)	Economic/strategic	Doesn't exist	Complete
Minimize/Substitute Hazard (MH)	Engineering	Human error plays a minor role	High
Physical Barriers (PB)			
Warning Devices (WD)	Assessing	Human error is possible	Medium
Procedures (P)	Administrative and work processes	Human error can play an important role	Low
Personnel Skills and Training (PST)			

If it is not possible to eliminate the hazard, attempts must be made to mitigate the potential effects of the hazard. Mitigation consists of actions to minimize the hazards (MH), most often with engineering methods, or to implement physical barriers (PB) capable of separating the hazard from the worker or the work process. Warning devices (WD) are often used to assess the performance of engineering controls (MH) and physical barriers (PB) or to prompt a change in administrative or work processes. Controls that are largely focused on operational and work process issues consist of procedures (P) and personnel skills and training (PST). Procedures (P) can often rely on the personnel skills and training (PST) of the worker. The reliance on worker behavior increases the potential for human error and reduces the risk reduction effectiveness when compared to mitigation efforts (*Table 11*). If controls fail to prevent the unwanted event or are not possible, then the hazard is tacitly tolerated and recovery measures are put into place to minimize losses. It is then appropriate to consider the hazard as a real threat and not just a potential threat. Recovery measures can include all control categories listed in *Table 11*.

Analyses of the ten case studies presented in this report require an evaluation of the controls identified by the risk assessment teams. These controls are the principal output of the MHRA. To accommodate this analysis, every identified control was assigned to one of the categories listed in *Table 11*. The characteristics of each category are given below:

Eliminate Hazard (EH) – The characteristics of this category are self-evident – elimination of the hazard under consideration. This can also be done with changes in equipment, changes to the mining process or method, or changes in the location of the hazard which eliminate personnel exposure.

Minimize or Substitue Hazard (MH) – The characteristics of this category consist mainly of engineering controls, i.e., improved ventilation, fire fighting equipment, backup systems, fire suppression systems, use of an event simulator, enhanced information about the hazard, improved construction / drilling / exploration techniques, electrical component performance characteristics and fault protection, designing to standards, improved equipment (values, brakes, tubing, etc.), available medical and rescue teams, etc.

Physical Barriers (PB) – The characteristics of this category are focused on physical barriers that separate the hazard from the worker, i.e., roof rock reinforcement, equipment skirting and guarding, sealing, rock dusting, refuge chambers, shielding, barriers, walls, special containers, heat wraps, self-rescuers, personal protective gear, etc.

Warning Devices (WD) – This category is primarily concerned with systems that monitor environmental / equipment conditions, i.e., gas monitors, PEDs, sampling pumps, gages, extensometers, tags, indicators, microseismic monitors, bag samples, alarms, sirens, certain kinds of communication systems, etc.

Procedures (P) – The characteristics of this category concentrate on processes conducted by workers and management, i.e., policies, inspections, checks, documentation, methods, roles, definition, restrictions, audits, purchases, investigations, standards, trigger action response plans (TARP), duties, work orders, updates specifications, process requirements, etc.

Personnel Skills and Training (PST) – The characteristics of this category center on training needs, personnel needs, required competency, testing, estimate consequences, reinforcing skills, mentoring, communication, expertise, behavior controls, operator errors, operator sensors, inspection quality, observation of conditions, introductions, clarification, etc.

An MHRA risk assessment team should strive for the high end of the hierarchy of controls. Some attempt must be made to at least consider how the hazard might be eliminated. This is often most easily accomplished in the early stages of a mining project's life cycle. Most often, the controls identified during the MHRA attempt to mitigate the hazard or to tolerate the hazard by putting into place recovery measures that will minimize losses. The team should be cautious of an over-reliance on warning devices that require manual readings, administrative procedures, and the personnel skills and training of the work force. In general, an MHRA should strive to go beyond the standards and regulations requirements for mining.

4.5 – Audit and Review

After an MHRA, a re-assessment of the site's hazards and an evaluation of the implemented risk mitigation program should be done on a regular basis by skilled and experienced personnel. It is also appropriate to audit and review the MHRA when rapid changes occur in some relevant work process or operational factor, i.e. design, construction, etc. In these cases, the audit and review can focus on the part or condition that is actually undergoing the change. An audit and review should, at minimum, determine the status of the risk management plan and make recommendations for improving potential deficiencies in the plan. Tools, such as a risk register, are sometimes used to help with auditing and reviewing important controls at a mining operation.

5.0 - MHRA Pilot Studies at US Underground Mining Operations

Ten case studies were performed at a wide cross-section of underground mines (*Table 12*). The mines make up important sectors of the US minerals industry, i.e. coal, metal, nonmetal and aggregate. The sizes of the mines ranged from small (< 50 miners) to large (>450 miners). Important mining states were represented in the pilot project including Pennsylvania, Ohio, West Virginia, Colorado, Montana, New Mexico and Alaska.

Table 12 - Characteristics of the 10 MHRA case study sites.

ID	Section	Size	Duration, days	Commodity	Risk Assessment Topic
A	5.1	Large	1.5	Metal	Rock reinforcement process
B	5.2	Small	0.3	Stone	Unplanned detonation of a production blast
C	5.3	Large	2	Coal	Spontaneous combustion causing fire/explosion
D	5.4	Large	2	Coal	Underground workshop fire
E	5.5	Small	1	Coal	Water inundation
F	5.6	Small	1	Stone	Escapeway egress blockage
G	5.7	Medium	2	Nonmetal	Natural gas ingress
H	5.8	Medium	3	Coal	Conveyor belt fire
I	5.9	Large	2	Coal	Longwall gate entry track fire
J	5.10	Large	4	Metal	Captive cut and fill change of mining method

These case studies document the MHRA performed at underground mines and discuss how risks for multiple fatality events were potentially reduced by reinforcing existing practices and processes and by adding new prevention controls and recovery measures. The common objective of these MHRAs is to 1) identify potential hazards that could cause a multiple fatality event, 2) determine which unwanted events pose the greatest threat for the mine, and 3) identify a potential plan to prevent the threat or recover from the consequence of the event happening. The controls typically consist of a broad spectrum of prevention, monitoring, first response, and emergency response techniques and help to move an operation from a reactive to a proactive approach towards safety.

Cooperating mining operations were typically selected after they first participated in a 3-day Workshop on Minerals Industry Risk Management. This workshop was organized by NIOSH and taught by Professor Jim Joy of the University of Queensland, the primary MISHC participant in the pilot project. Two workshops occurred, August 3-5, 2006, in Spokane, Washington, and December 18-20, 2006, in Pittsburgh, Pennsylvania. In many cases, the cooperating mining operations selected the hazard to be examined during the case study during, or shortly after attending, the above workshops.

5.1 – Rock Reinforcement Process Risk Assessment Case Study

Mine A is a relatively new operation located in a remote area. This operation is using a mechanized cut-and-fill method to mine a vein deposit. After the ore is blasted and mucked at underground faces, it is trucked to a surface processing facility for gold recovery using gravity, chemical and pyro-metallurgical processes.

Representatives from the mine's parent company attended a NIOSH-sponsored MHRA training course where two risk assessment scoping topics were identified.

5.1.1 - Risk Assessment Scope

Two risk assessment scopes were originally proposed:
- Mine Fire Risk Assessment - 1) review hazards associated with the potential for an equipment fire in the intake air stream, 2) evaluate strategies and techniques for early detection of the hazard, and 3) evaluate the escape and emergency response plan for a major fire at this mine.
- Dissolved Oxygen Plant Risk Assessment - 1) review hazards associated with the dissolved oxygen plant, 2) evaluate strategies and techniques for early detection of the hazard, and 3) evaluate the escape and emergency response plan for the major hazard and impact on other site facilities.

5.1.2 – Risk Assessment Team

Two teams were formed by the cooperator and were designated to spend one day on each topic. The first team represented underground mine operations and included:
    Mine manager
    Mobile maintenance foreman
    Miner
    Mine operations trainer
    Safety coordinator
    Safety/human resources manager
    NIOSH observer
    Facilitator – MISHC (University of Queensland)

The second team included processing plant personnel:
    Assistant mine manager
    Processing plant safety trainer
    Maintenance engineer
    Maintenance foreman
    Maintenance superintendent
    Plant labourer
    Mechanic
    Electrician
    NIOSH observers
    Facilitator – MISHC (University of Queensland)

Both teams had representatives from a wide cross-section of the mine. Each team pursued an independent risk assessment.

5.1.3 - Structure of the Risk Assessment:

After the initial meeting of the mine's appointed risk assessment team, it became apparent that the team lacked adequate training in fundamental risk management principles and were unfamiliar with risk assessment techniques and tools. In addition, the risk assessment team did not have the necessary expertise to sufficiently analyze the process identified in the risk assessment scope. Without an understanding of risk assessment techniques and tools and a lack of knowledge about the processes to be analyzed, it was decided that an MHRA could not be performed on the suggested topics. The team decided instead to focus on a general hazard identification and job/process mapping discussion aimed at introducing risk assessment techniques to the operations.

To maximize the training potential associated with this case study, half of each day was dedicated to instruction/training and the other half to developing a JSA for a key process. The underground and surface teams were made up of management, front line supervisors, tradespersons, engineers and labor. The development of JSAs reinforced the concepts of controlling energies through design of appropriate work processes.

5.1.4 – New Risk Assessment Topics:

After the instructional aspects of the risk assessment were completed, both teams identified a critical safety process that could be mapped and examined by the WRAC tool. The underground operations team decided to examine the process associated with the selection and installation of rock reinforcement. The process plant operations team selected repair of a frequently failing slurry pump. Neither of these potential hazards represented a high-consequence event that might lead to a multiple fatality accident, but they did represent problems relevant and instructive to the case study participants.

5.1.4.1 - Rock Reinforcement Process:

Mine A uses an automated drill to install rock reinforcement (*Figure 10*). The underground operations team was able to rapidly describe the steps in the selection and installation of rock reinforcement, as follows:
1. Scale
2. Select bolt size and type
3. Select bolt pattern
4. Stock machine with supplies
5. Move machine into heading and set-up
6. Drill back hole
7. Insert bolt
8. Anchor wire mesh screen (repeat steps 6, 7 and 8 until ring is complete)
9. Advance to next ring (start at step 6)

Figure 10 - Drill used to install rock reinforcement.

The team then attempted to identify the hazards for each step in the rock reinforcement process and rank the associated risks. This was accomplished with a WRAC. Every potential unwanted event was examined and their Maximum Reasonable Consequence (MRC) and likelihood of occurrence were determined. Due to time constraints, only hazards associated with steps 1 through 5 were examined during the exercise. These steps included scaling (step 1), the selection of bolts and patterns by the miner (steps 2 and 3), and the set-up (steps 4 and 5) of the machine by the miner, i.e. choosing drill bit sizes, etc. (*Table 13*). Risks were ranked using a 5x5 matrix (*Table 14*).

Table 13 - WRAC of the initial steps in the rock reinforcement process.

Step in Process	Unwanted Event	MRC	Likelihood	Rank
1. Scale fresh ground	Person in poor position	C	4	9
	Bar too heavy	C	2	17
	Ground not scaled	B	4	5
	Poor quality scaling	B	4	5
2. Select bolt type & size	Wrong bolt selected for conditions	B	4	5
	Correct bolt unavailable, substitute bolt inadequate for pattern	A	4	2
	Short bolts installed intentionally	B	4	5
3. Select Pattern	Wrong pattern selected for conditions	B	4	5
4. Stock Machine	Improper lifting of materials	C	3	13
5. Machine Set-up	Incorrect bit size selected for bolt	B	4	5
	Machine positioned too far forward	B	3	8
	Moving energized cable by hand	B	1	16

Table 14 - A 5x5 risk matrix used to rank risk in the rock reinforcement process WRAC.

Maximum Reasonable Consequence	Likelihood of Occurrence				
	5-Common (>1/week)	4-Likely (1/month)	3-Moderate (1/year)	2-Unlikely (1/several years)	1-Very Unlikely (almost never)
A-Multiple Fatality	1	2	4	7	11
B-Fatality	3	5	8	12	16
C-Lost-time Injury	6	9	13	17	20
D-Reportable Injury	10	14	18	21	23
E-First Aid Injury	15	19	22	24	25

The results of this partial WRAC were then subjected to a Bow Tie Analysis (Figure 8). The highest unwanted events from the WRAC (*Table 13*) formed the center of the bow tie. The reliance on the judgement of the miner to manage these five critical steps was identified as a key vulnerability by the risk assessment team.

The risk assessment team then identified existing and new controls to manage these top unwanted events, i.e. the left side of the BTA. For example, the team identified the use of long (14-ft) scaling bars and the training and experience of the miners as the existing controls. In addition, several new controls were suggested by the team to increase the training effort associated with the rock reinforcement process (*Table 15*). In particular, the risk assessment team recommended that front line supervisors should monitor the rock reinforcement process by the miners on a daily basis.

Table 15 - New ideas for preventing rock reinforcement selection and installation failures.

New prevention ideas	NI1	Improve training and monitoring of miners in selection of bolts and patterns
	NI2	Improve training and monitoring of miners in proper set-up of machine
	NI3	Improve training of miners in use of scaling bars, only use 14-ft bars
	NI4	Improve training in lifting and keep material storage area clean

NI = New Ideas

Time constraints prohibited a complete analysis of the remaining steps in the rock reinforcement cycle. It was understood that the group would continue the risk assessment on their own, examining the other steps in the rock reinforcement process in a similar manner. At the end of this process, a complete list of existing controls and recovery measures would be compiled and the new ideas examined to produce a Rock Reinforcement JSA. This JSA would be used to support the training of those involved in the rock reinforcement process and monitor their performance.

5.1.4.2 - Slurry Pump Repair Process:

After the morning training session the surface operations team mapped the process for repairing the slurry pump:
1. Notify control room
2. Lockout pump valves
3. Secure connecting values
4. Bleed-off system pressure

5. Remove pump housing
6. Replace failed components
7. Reinstall housing
8. Remove locks and open valves
9. Restart

The team struggled with identifying the steps in this process and it became clear there were multiple methods being used to secure the valves and drain the pressure from the pump systems. Once there was general agreement on the steps in the process, discussions on the hazards associated with steps 1 through 4 were initiated. However, a continued lack of consensus concerning proper procedures to control the hazards in these steps limited progress within the allotted time. The team recognized the repair process as an elevated risk activity because of its high likelihood of occurrence. The team also agreed that the mine should develop an SOP for this reoccurring activity or investigate alternative equipment that would not require such frequent repairs. Despite this recognition, the team did not feel empowered to effect this change without further approval from management.

5.1.5 - Discuss Implementation, Monitoring and Auditing Issues:

The overall acceptance and understanding of the risk assessment process by the team at this mine was limited. The team selected was energetic and eager to learn the subject but had limited exposure to risk assessment practices and as such does not utilize low level risk assessment tools, such as JSAs or SOPs, in their mining processes.

The management team at Mine A was short on staff and expressed a reluctance to take on additional administrative burdens associated with the MHRA process. The limited time available for the MHRA exercise was largely due to existing workload obstacles. This is a new mining operation with a relatively high turnover rate where site-specific experience is in relatively short supply. In its current state, it is difficult to see how management could supply the necessary guidance to implement, monitor and audit a formal risk management approach such as an MHRA.

Lastly, it was not clear that labor and management were communicating at a level necessary to foster the exchange of ideas in a frank and productive fashion required for an MHRA. While there was a cordial labor-management relationship and communication was good for day-to-day operational issues, there seemed to be a rigid separation between labor and management in the strategic decision-making processes. Future involvement of the workforce in development of JSAs and SOPs may help to improve this relationship and better position the mine for effective major hazard management planning.

5.2 – Unplanned Detonation of a Production Blast Risk Assessment Case Study

Mine B is an underground limestone mine using the room-and-pillar mining method. Rooms are typically 35 ft wide by 20 ft high. The production faces are drilled with a V-cut pattern 13 ft deep and filled with ANFO. Blasting caps and primers are inserted into each hole and connected to a specific timing delay. Typically the faces are drilled and loaded early in the shift. Faces prepared in this manner are barricaded and ready to connect to the blasters via prima cord at the shift's end. Leaving the face fully charged represents a hazard for miners working near these areas.

A team of company and external personnel took part in a systematic discussion about the explosives hazards. The team focused on work processes associated with blast hole loading and detonation and specific products used in the blasting process. The cooperating company identified the unplanned detonation hazard after attending the previously discussed NIOSH sponsored workshop on Minerals Industry Risk Management.

5.2.1 - Risk Assessment Scope

The objective of the risk assessment was to 1) review hazards associated with unplanned blasthole detonations prior to scheduled face shots at Mine B, and 2) evaluate strategies and techniques to mitigate the risk.

5.2.2 - Risk Assessment Team

The team was made up of persons employed at the mine, technical representatives of the local explosives supplier, and an observer and facilitator. More specifically, the team members included:
    Two management representatives
    One engineer
    One miner
    One underground mine foreman
    Two external explosives experts
    NIOSH observer
    Facilitator – MISHC (University of Queensland)

5.2.3 – Risk Assessment

A formal risk assessment method was not completed, nor was it needed to solve the unplanned detonation events. The risk assessment was largely accomplished through focused discussions of the hazards and potential controls. The entire process took approximately 2 hours. For the purposes of this analysis, the various discussion segments were placed into the steps associated with the MHRA process (see below).

5.2.3.1 – Step 1, Identify and Characterize Unplanned Detonation Mining Hazards

The primary safety risk identified was the unplanned detonation of a production face while mining personnel were present. The risks were the greatest when the blast holes were loaded with explosives and all the blasting caps were connected together. The risks were increased by the length of time the face sat in this condition. At this mine site, that length of time ranged from 4 to 6 hours. The team agreed that the risks could be significantly reduced if the length of time between setting up the face for a blast and actually blasting the face was shortened.

The discussions began with a representative of the mining operation describing the work processes used in setting up a development face for a production blast. This was followed by a technical presentation from representatives of the explosive manufacturer. The focus of this discussion was on the products used in the blasting process.

5.2.3.2 – Step 2, Rank Potential Unwanted Events

Numerous potential threats were identified capable of causing a spontaneous ignition event:
1. *Lightning strike on the surface above the underground mine* – Understandably, blasts are not attempted in surface operations when lightning occurs. However, in underground mines, blasting procedures are not generally altered when lightning exists on the surface. Recently, unplanned detonations have occurred when lightning strikes were observed near the blasting area (Anon, 2005). Also, the association of lightning with the Sago Mine Disaster in January, 2006 (Gates et al., 2007), has provided a potential example where electrical currents may pass deep within the earth's strata in conjunction with surface lightning strikes. At Mine B, there were no procedures that altered the blasting practices during weather that produced lightning strikes.
2. *Rock fall hitting base charge detonator and the strike initiates an electric spark* – A rock falling from the mine's roof and directly impacting a base charge detonator could initiate an electric spark and cause an unwanted detonation of the explosives. At Mine B, precautions are taken to remove all loose material from the roof prior to loading the blast holes. Both mechanical and hand-scaling techniques are used.
3. *Shovel or other tools impacting the detonator* – This is a similar problem to the roof rocks hitting the base charge detonator. The potential for this type of impact detonation was considered to be very low.
4. *Static electric charge* - Most electric detonators have protective systems built in at the manufacturing stage to eliminate unplanned detonations. However, when static electricity is known to be a severe problem, such as in dry or dusty conditions, then particular care is needed to prevent an accidental discharge to a circuit. Sometimes mining operations require the miner to wear conducting footwear as a precaution. No special gear was worn at the Mine B site.
5. *Snap and shoot* – Several accidents have occurred in the last fifteen years related to snap and shoot[4]. Holmberg and Salomonsson (2002) identified five accidents in which a shock tube might have been unintentionally stretched to breakage. It was believed that in some instances this caused the tubes to initiate. For example, if a vehicle ran over a blast hole with a shock tube, it is possible that the tube could be entangled in the vehicle and be caused to snap and shoot.

The snap and shoot threat was considered to be the most likely scenario for an ignition of a

---

[4] Snap and shoot is sometimes also called stretch and shoot; snap, slap and shoot; or whip, snap and shoot.

face loaded with explosives at the Mine B site. Although the other four threats were all potentially possible, the team felt there were sufficient barriers and controls currently in place to minimize their likelihood. Therefore, the focus of the team was to examine what might be done to reduce the snap and shot threat.

5.2.3.3 – Step 3, Determine Important Existing Prevention Controls and Recovery Measures

The existing work process associated with the snap and shot threat was to:
1. Place the blast caps and ANFO into the blast holes early in the shift
2. Tie the blasting caps and shock cord (*Figure 11*) in place early in the shift, and
3. Connect these items to the detonation system at the end of the shift, just prior to blasting.

The time between 2 and 3 was between 4 and 6 hours. During that time, there existed an opportunity for an unplanned detonation through the snap and shoot mechanism.

Figure 11 - Photograph of shock cords similar to the ones used at the study site.

5.2.3.4 - Step 4, Identify New Prevention Controls and Recovery Measures

The team discussed two options to reduce the risk for snap and shoot:
1. Use the current work procedure to load and blast the face and attempt to decrease the likelihood of unplanned detonation through new prevention controls and recovery measures.
2. Change the work procedure to tie on the caps just before the blast, thereby eliminating the threat of premature ignition (EH).

As the team discussed the hazard in more detail, it became apparent that a formal risk analysis technique was not needed to identify an appropriate safety solution. The team agreed to recommend altering work procedures so that final blast set-ups were done at the end of shift (option 2). In many ways this process followed an informal fault tree analysis where the process was altered to eliminate paths to the top event.

5.2.3.5 – Step 5, Discuss Implementation, Monitoring and Auditing Issues

Finally, the changed steps were reviewed to identify any new risks. No significant new risks were identified so the team agreed to recommend that the procedure be changed. The entire process took less than 2 hours to complete.

The team made a single effective recommendation, as follows:
> The blast caps should not be snapped in place when the face is loaded and tied in early in the shift. During the morning blast setup work, the blast caps should be located near the face but at least 2 feet from the setup. The shot firer should return at the end of shift, when he normally inspects the blast setup, and tie-in the caps at that point. The shot firer should subsequently inspect the set-up and, if appropriate, proceed with the usual final blast setup and detonation. This recommendation should be part of a Standard Operating Procedure for connecting the detonation systems to loaded blast holes at this operation.

This case study did not apply a formal risk assessment method but rather undertook a structured discussion of the hazard. In this case study, new controls were not needed since altering the work procedure eliminated the hazard and hence the risk. This informal risk assessment process appeared to be an effective and valued process by the team members.

5.3 – Spontaneous Combustion Causing Fire/Explosion Risk Assessment Case Study

Mine C is a deep longwall coal mine with the potential for spontaneous combustion. The mine uses a bleederless ventilation system in its longwall panels to help reduce the potential for spontaneous combustion. Spontaneous combustion can occur in coalbeds with physical and chemical characteristics that allow the coal to oxidize at relatively low temperatures. This can set into motion a chain of events where additional oxidation causes temperatures to rise to a point where a flame will occur. A primary ingredient for this reaction is oxygen. If oxygen is not present, the oxidation process cannot occur. Bleederless ventilation systems have been extensively used in many parts of the world as a spontaneous combustion control measure (Smith et al., 1994). These systems are designed to reduce oxygen contents in areas where the coal has been extracted through isolation from a mine's ventilation system. Smith et al. (1994) reported the two areas that provided the most risk for spontaneous combustion are those around the seals and directly behind the longwall shield supports. MSHA's regulations covering bleederless ventilation systems are found in Federal Code of Regulations under Part 30, Section 75.334 (f).

The bleederless ventilation system at Mine C uses seals at gate entry cross-cuts to separate ventilating air from the gob. The mine's greatest concern is the potential for spontaneous combustion in the gob area behind the active longwall face. A risk assessment was performed to investigate major hazard potentials and evaluate controls to mitigate the potential for a spontaneous combustion event.

5.3.1 - Risk Assessment Scope

The scope of the risk assessment was originally designed to utilize a WRAC to review the hazards and a BTA to identify existing and new prevention controls and recovery measures from the spontaneous combustion hazard. Mine management did not attend a NIOSH-sponsored MHRA training course prior to the site study. At the start of the Risk Assessment, the team was briefed on risk management principles. Reference materials on sources of spontaneous combustion were used to provide a detailed list of relevant energies in need of control. Unfortunately, the risk assessment was limited to a single day.

5.3.2 - Risk Assessment Team

The team consisted of mine site personnel directly involved in monitoring and responding to a spontaneous combustion event, a NIOSH observer, and a facilitator from MISHC. One member of the general workforce was represented on the risk assessment team. Team members were as follows:
- Mine operations manager
- Technical services manager
- General mine foreman
- Safety manager (partial)
- Fireboss (labor)
- NIOSH observer
- Facilitator – MISHC (University of Queensland)

5.3.3 - Risk Assessment

MSHA statistics show that spontaneous combustion accounted for 17% of the ignition sources for the 87 reportable fires occurring at underground coal mines between 1990 and 2000. The potential for the spontaneous combustion hazard increasing the risk of fire and explosion was a primary concern at the case study mine. In the past, the mine has routinely sealed gob areas. No reportable fires have occurred at the mine according the MSHA database. However, spontaneous combustion events have occurred at nearby mines operating in the same seam.

Currently, Mine C uses a modified "U" ventilation system where the air is brought up the headgate entries, across the face, and down the tailgate entry (*Figure 12*). In the future, the operator is planning to change the manner in which the gob will be sealed. The plan is to temporarily retain bleeder headings around the gobs while development continues inby the panel. As a result of this change and the historical occurrence of this hazard in adjacent mines, the mine decided to examine risks related to spontaneous combustion in the active panel gob.

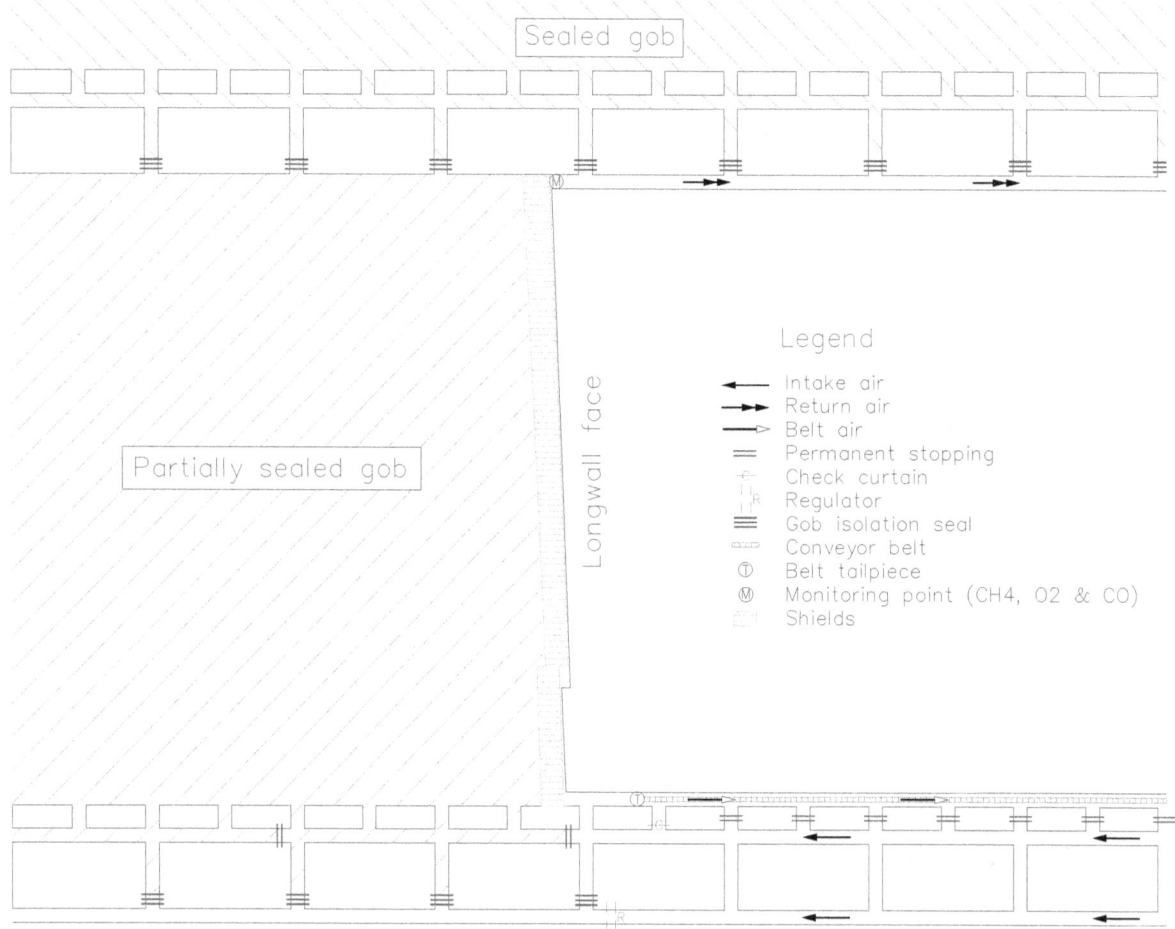

Figure 12 - Bleederless ventilation system used at the study mine to control spontaneous combustion.

Following a briefing on the principles of risk management and the tools to be used, the facilitator led the team through a structured process involving the following steps:

1. Recognizing and characterizing hazards
2. Selecting top unwanted events (informal risk ranking)
3. Identifying existing prevention controls and recovery measures with the BTA (partial list)
4. Identifying new prevention controls and recovery measure ideas with the BTA (partial list)
5. Implementation, monitoring and auditing issues.

Because of the time constraints, it was not possible to formally risk rank the top unwanted events or to fully develop the list of existing or new prevention controls and recovery measures.

5.3.3.1 – Step 1, Identify and Characterize Spontaneous Combustion Hazards in the Active Panel Gob

Spontaneous combustion hazards within current and planned longwall panels are influenced by potential heat sources and conditions of the atmosphere in the gob (*Table 16*). The type of heat sources in or near the gob in Mine C included oxidation of the broken coal, hot works in the gate entries or along the longwall face, sparks from roof bolts failing as the strata collapsed in the gob, and heating of an overlying coalbed from an igneous intrusion. Conditions of the atmosphere in the gob needed to cause a fire or an explosion include availability of oxygen from the longwall face ventilation and hydrocarbons from the mined coalbed or from the overlying sandstone and coalbed.

Table 16 – Potential heat sources and conditions of the atmosphere in the gob needed to cause a fire or an explosion.

Heat Sources	Gob Atmospheres
Coal oxidation (spontaneous combustion)	Oxygen available from face ventilation
Welding/cutting (hot works)	Hydrocarbon liberation from coal
Roof bolt sparks during gob caving	Gas migration from sandstone main roof
Heat in upper coal due to igneous intrusions	Gas migration from the upper coalbed as caving occurs

5.3.3.2 – Step 2, Rank Potential Unwanted Events

Time constraints prohibited the risk assessment team from a thorough discussion of all potential unwanted events that could lead to a fire or explosion. The team identified three main categories of events that could potentially produce spontaneous combustion in the gob of an active longwall panel:
1. Air (oxygen) flow into the gob through
    a. Headings / cross-cuts (including barometric pressure change effects)
    b. Gob boreholes
    c. Longwall shield supports
    d. Leaky seals
2. Methane / hydrocarbon buildup in the gob to ignitable levels (5 to 15 %) through
    a. Gas migration from overlying coalbed into gob
    b. Gas migration from overlying sandstone into gob
3. Heat source in gob where broken coal in gob from caving upper seam increases oxidation rates

The team selected these unwanted events based on the degree to which the most workers might be exposed and on the historical information concerning past spontaneous combustion episodes. In the absence of any formal risk ranking method such as the WRAC or the PHA, the risk assessment team informally agreed that all unwanted events were of high consequence to the mining operation.

5.3.3.3 – Step 3, Determine Important Existing Prevention Controls and Recovery Measures

A BTA was performed on only the four events under the air flow into the gob category due to time limitations. The risk assessment team discussed potential connections between the hazards listed in *Table 16* and the mining operations. The team then went on to identify 17 existing prevention controls (*Table 17*).

Table 17 - Existing key prevention controls for the spontaneous combustion risk assessment.

TOP EVENT => Air (Oxygen) Flow into the Gob		
Through headings / cross-cuts (including barometric pressure change effects)	PC1	Forced Air Ventilation that causes positive pressure on gob (MH)
	PC2	High-quality seal design with ring grouting to improve seal effectiveness and reduce roof-to-floor convergence (PB)
	PC3	Location of seals to avoid leakage pathways (MH)
	PC4	Seals checked and maintained by fire boss weekly (P)
	PC5	Monitoring pressure balance across seals weekly (WD)
	PC6	Atmosphere behind seals tested weekly, gas tested by experienced person (WD)
	PC7	Gob vent borehole gasses tested weekly, gas tested by experienced person (WD)
Through gob boreholes (including barometric pressure effects)	PC8	Forced air ventilation causes vent holes to only breathe out (MH)
	PC9	Gob hole is sealed when panel is complete and sealed (PB)
	PC10	Gob holes can be shut in if oxygen >10% (P)
	PC11	Check valve & flame arrestors assure no flow down holes into the gob (MH)
Through longwall shield supports	PC12	Curtains on shields at start up (PB)
	PC13	Monitoring of gob to verify it is tight (P)
	PC14	Real time gas monitoring at tailgate to detect heating (WD)
	PC15	Seals installed in gates to reduce flow into gob area from gates (MH)
Through leaky seals	PC16	Seals tested weekly with smoke tube and repaired as needed (P)
	PC17	Flexible foam packs used for construction and repair (MH)
	As above: see PC2	
TOP EVENT => Methane / hydrocarbon build up in the gob to ignitable levels (5 to 15 %)		
Gas migration from overlying coalbed into gob	Not addressed due to time constraints	
Gas migration from overlying sandstone into gob	Not addressed due to time constraints	
TOP EVENT => Heat source in gob		
Broken coal in gob from caving upper seam increasing oxidation rates	Not addressed due to time constraints	

PC – Prevention Controls
MH – Minimize Hazard
PB – Physical Barrier
WD – Warning Devices
P – Procedures
PST – Personnel Skills and Training

The 17 prevention controls were distributed between the minimize hazards (MH), physical barriers (PB), warning devices (WD), and procedures (P) control categories (*Figure 13*). Eliminating the hazard (EH) could only be done by not mining the coalbed with conventional mining techniques, since this coal is prone to spontaneous combustion. Many of the controls were required by MSHA regulations while others were considered to be Best Practice at this mine site.

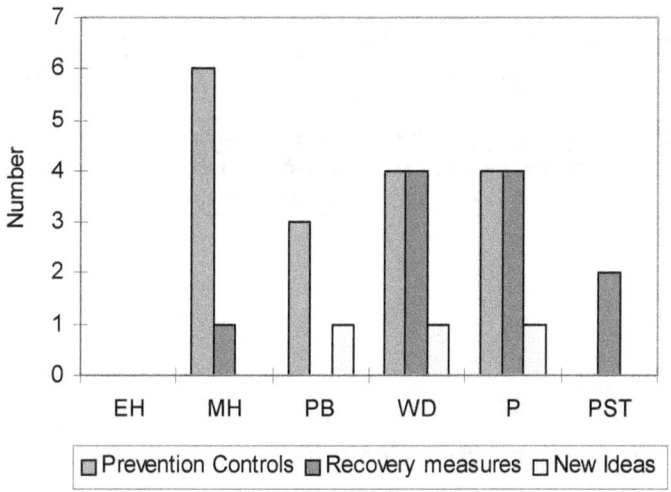

Figure 13 - Distribution of prevention controls and recovery measures for the spontaneous combustion causing fire/explosion risk assessments.

The team identified 11 recovery measures designed to mitigate the consequence of a spontaneous combustion event (*Table 18*). Discussions concentrated on monitoring and initial evacuation responses (*Table 18*). These existing recovery measures are mostly aimed at complying with MSHA standards and regulations for coal mines with spontaneous combustion hazards. Many should be documented in the mine's ventilation plan. A common theme of the discussions of these controls was the dependence on a few individuals to ensure compliance, where 72% of the recovery measures (*Figure 13*) were either warning devices (WD) or procedures (P). Formal auditing methods were not always used for critical controls. The mine management at Mine C has experience in reacting to heating events and seems to rely on leadership skills rather than formal procedures for directing response.

Table 18 – Existing key recovery measures for the spontaneous combustion risk assessment.

Early Detection		
Advancing spontaneous combustion in the gob	RM1	Weekly bag sampling at seals every 1000 ft along seal line/perimeter (WD)
	RM2	An on-site Gas Chromatograph with backup at other mines in the valley (WD)
	RM3	Written Action Levels/Trigger Points; (WD) 　　Level 1: >5ppm H2 AND >100 ppm CO triggers 　　　　If $O_2 > 5\%$ resample ASAP 　　　　If $O_2 < 5\%$ resample within 24 hrs 　　Level 2: >5ppm H2 AND >100 ppm CO on resample triggers 　　　　If $O_2 > 10\%$ resample every 4 hrs 　　　　If $O_2 < 10\%$ resample within 8 hrs 　　Level 3: >5% H2 OR >350 ppm CO OR rising CO/CO_2 ratio 　　　　AND if $O_2 > 10\%$ resample every 3 hrs and notify MSHA

		Level 4: >1500 ppm CO OR HC>4% AND O_2>10% then evacuate all non-essential persons from mine and call MSHA
Fire detected in the gob and initial response (communicate gob fire to miners and specify the correct egress path)	RM4	Mine dispatch and shift foreman are instructed to evacuate mine by phone (P)
	RM5	Alarms at belt feeders and mine pager system that indicate need to egress in both this mine and the overlying seam (WD)
	RM6	Persons entering remote areas are required to notify dispatch prior to entering. Dispatch keeps written record of personnel locations and entry/exit time (P)
	RM7	Shift foreman aware of persons assigned to tasks in remote areas (P)
Evacuation		
Rapid egress from the mine required	RM8	Regular Emergency Response training of all underground personnel (PST)
	RM9	Live training to practice egress during likely event scenarios (PST)
Explosion Prevention		
Fire event escalates to explosion	RM10	Rock dusting of mine exceeds minimum standards (MH)
	RM11	Pre-framing of fire seals across longwall gate roads and staging of materials nearby for rapid sealing of involved area (P)

RM – Recovery Measures
MH – Minimize Hazard
PB – Physical Barrier
WD – Warning Devices
P – Procedures
PST – Personnel Skills and Training

5.3.3.4 - Step 4, Identify New Prevention Controls and Recovery Measures

As part of the BTA, new prevention control and recovery measure ideas were identified by the team. However, the limited time available with the mine staff and the need to begin the process with basic training in the risk management process limited the extent to which the exercise could be completed. The team identified one new prevention control and two new recovery measure ideas (*Table 19*). Mine management agreed to review these new ideas and decide if they should be implemented.

Table 19 - New ideas for mitigating risk of spontaneous combustion.

Prevention Controls	
NI1	Improve rock dusting of gob perimeter such as bottom dusting (PB)
Recovery Measures	
NI2	Investigate stench gas or other methods of increasing the likelihood that personnel in remote areas get the message to egress (WD)
NI3	Develop a detailed vendor list for materials, equipment and expertise for use in a spontaneous combustion emergency response (P)

NI – New Ideas
PB – Physical Barrier
WD – Warning Devices
P – Procedures

5.3.3.5 – Step 5, Discussion of Implementation, Monitoring and Auditing Issues

The team identified 17 existing controls for prevention of spontaneous combustion in the gob; most (8) can be considered engineering controls (MH and PB), 4 are manual monitoring controls (WD), and 5 are process controls. The 6 response measures are primarily administrative (P and PST) with 4 manual monitoring (WD) and one engineering control (MH).

Influencing the number and type of controls identified is the mine's general avoidance of formal procedures other than those mandated by regulations. Mine management expressed a preference for flexibility in actions taken based on information collected through the multiple manual monitoring controls and the experience of the staff.

The overall acceptance and understanding of the risk assessment process by the team at Mine C was limited. The team selected was energetic and knowledgeable in the subject hazard but had limited exposure to risk assessment practices. The mine does not utilize informal risk assessment tools such as JSAs or SOPs in its mining processes. The management team at this mine carries multiple responsibilities and expressed a reluctance to take on more administrative burdens. This existing workload obstacle was evident in the limited time available for the exercise.

One issue with implementing the risk management principles is the willingness of the team to rely on manual monitoring of the hazard and to defer most actions to the judgment of a few staff members. While these experienced staff members would likely act appropriately in monitoring for hazards, this system exposes the mine to delays should events begin when key staff are not accessible.

One other aspect of risk management not completely addressed by the team was that the formal response plans end with evacuation of the mine and notification of MSHA. While it was clear that the mine would indeed take actions to combat the fire and recover the mine, these response plans would be developed by key staff spontaneously in reaction to unfolding events. In these respects the outcomes of this exercise may be the product of a cultural acceptance of relatively high levels of risk at Mine C.

5.4 – Underground Workshop Fire Risk Assessment Case Study

Mine D is a large longwall coal mine with over 600 salaried and hourly employees. The mine has an underground workshop located close to the bottom of the intake shaft (*Figure 14*). The workshop is basically a heading containing a pit large enough to allow maintenance work under a locomotive. Other workshop-related tasks are also performed in the immediate area such as battery charging. An underground workshop fire is considered to be a major hazard since it could potentially disrupt normal escapeway egress routes via the intake air shaft.

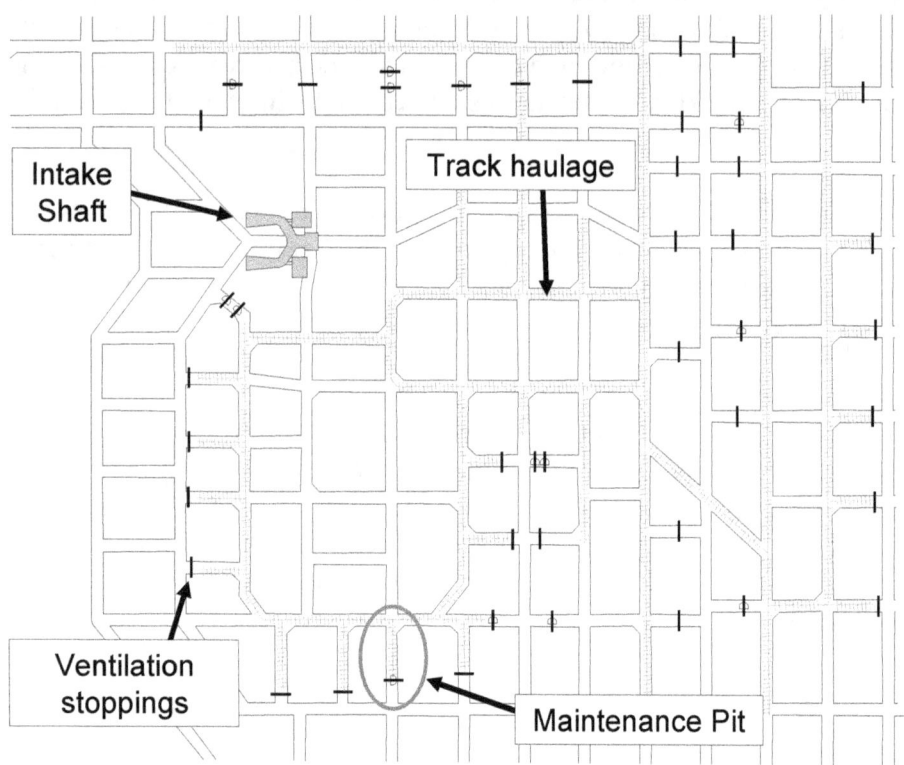

Figure 14 - Map showing the location of the maintenance pit with respect to track haulage, ventilation stoppings, and intake shaft.

The risk assessment project theme was identified at the NIOSH-sponsored workshop on Minerals Industry Risk Management and further developed through discussion between management personnel and NIOSH representatives.

5.4.1 - Risk Assessment Scope

The objective of this risk assessment was to 1) review fire hazards in the underground workshop located close to the mine's intake shaft, and 2) evaluate strategies and techniques to mitigate the risks.

5.4.2 - Risk Assessment Team

The team was made up of persons employed at the mine and by the mine's parent company, two external fire prevention experts, a NIOSH observer, and a facilitator from MISHC. More

specifically, the team members included the following:
- Four representatives from the company's safety program
- A general maintenance mine foreman
- A continuous improvement coordinator
- A fire brigade supervisor
- Two external fire prevention experts (NIOSH / Insurance Company)
- NIOSH observer
- Facilitator – MISHC (University of Queensland)

5.4.3 – Risk Assessment

The Risk Assessment involved facilitation of a team of personnel through a structured process of the following steps:
1. Hazard descriptions
2. Preliminary Hazard Analysis (PHA) method introduction and risk ranking
3. Potential unwanted event identification using PHA
4. Bow Tie Analysis (BTA) method introduction
5. Causes and prevention controls discussion using BTA
6. Consequences and loss reduction controls discussion using BTA
7. Repeat of Steps 5 and 6 for all high ranked potential unwanted events identified in Step 3.

5.4.3.1 – Step 1, Identify and Characterize Fire Hazards in the Underground Workshop at the Bottom of the Intake Shaft

The first step in the risk assessment involved identifying and understanding the hazards related to a fire in the underground workshop. The team brainstormed the potential heat and fuel sources that might be either available in the workshop or related to the maintenance functions performed in the workshop (*Table 20*).

Table 20 - Fire hazards consisting of potential heat and fuel sources.

Heat Sources	Fuel sources
Electrical power	Oils/grease
Welding/cutting (hot works)	Paper/trash
Grinding	Coal
Portable heaters	Diesel fuel
Batteries	Solvents
Hot engines/surfaces	Plastic
Exhaust/DPM	Wood
Spontaneous Combustion from oily rags	Hoses
Compressors	Acetylene
	Hydrogen (batteries)

After discussing the above hazards, the team used its knowledge of the underground workshop design and operation to undertake a Preliminary Hazard Analysis (PHA). The PHA identified a list of fire-related hazards within and near the maintenance area and then prioritized these hazards through a risk ranking process.

5.4.3.2 – Step 2, Rank Potential Unwanted Events

The risk assessment team identified 24 potential incidents/accidents related to an underground workshop fire (*Table 21*). The PHA risk analysis tool was selected to risk rank the potential incidents/accidents. A discussion of the PHA technique can be found in Section 3.2.2.

Table 21 - Preliminary Hazard Assessment (PHA) form for the underground workshop fire risk assessment case study.

	Potential incident/accident	Exposure (*Table 6*)	Overall Likeli-hood (*Table 7*)	Max. Reasonable Consequence (*Table 3*)	Most Likely Conse-quence (*Table 8*)	Risk Rank (*Table 9*)
1	Pressurized/atomized hydraulic oil from equipment sprays on heat source causing fire	A	B	*B*	B	5
2	Short circuit in battery cells leads to fire	D	C	*B*	B	8
3	Fire in garbage storage	A	C	*C*	B	8
4	Exhaust DPM filter fire when filter withdrawn from engine	B	B	*D*	C	9
5	Hydrogen explosion during repairs leads to fire	E	E	*D*	A	11
6	Auto battery fault when charging leads to fire/explosion	D	D	*D*	B	12
7	Spontaneous from oil rags/trash	A	C	*D*	C	13
8	Overheated compressor	B	C	*D*	C	13
9	Welding methods cause fire	D	C	*C*	C	13
10	Hot slag contacts combustible leads to fire	C	C	*C*	C	13
11	New rescuers ignites when hit by/run over in a workshop	A	D	*B*	C	17
12	Short circuit/fault on diesel equipment	A	D	*C*	C	17
13	Diesel equipment overheats	A	D	*C*	C	17
14	Diesel fuel is vaporized by heat and ignited by fault/short	A	E	*B*	C	20
15	Damaged and worn electrical equipment leads to fire	D	C	*D*	E	22
16	Extension cord/wire run over/damaged leads to a fire	D	C	*E*	E	22
17	Overheated brakes on equipment	B	D	*E*	E	24
18	Broken light fault	A	D	*E*	E	24
19	Grinding sparks lead to fire	D	D	*D*	E	24
20	Heater fault/circuit	B	D	*E*	E	24
21	O_2 problem leads to more susceptible conditions for fire	C	D	*B*	E	24
22	Transformer fault leads to fire	A	D	*E*	E	24
23	Contaminants onto electric heaters	B	D	*E*	E	24
24	Arch welding cable short/fault leads to a fire	C	D	*D*	E	24

The team decided to focus on the top four ranked risks from the PHA analysis. These potential incidents/accidents are:
1. Pressurized/atomized hydraulic oil from equipment sprays on heat source causing fire
2. Short circuit in battery cell leads to fire
3. Fire in garbage storage
4. Exhaust DPM filter fire when filter withdrawn from engine.

In addition, once the team began to discuss existing prevention controls and recovery measures, a decision was made to add another potential incident/accident:
5. Pool of oil from a failed pump ignites in the maintenance pit and leads to a mine fire.

5.4.3.3 – Step 3, Determine Important Existing Prevention Controls and Recovery Measures

A BTA was performed on each of the four top ranked risks from the PHA with one additional risk added after the PHA by the team. For many of the top ranked incidents, multiple potential causes were possible. For example, the <u>pressurized/atomized hydraulic oil from equipment sprays on heat source incident</u> contained seven different potential causes:
1. Worn hydraulic hose or coupler failure and pressure release
2. Hot work damages hose
3. Unknowing removal of the hose under pressure (towed in or running)
4. Hose pinched during installation or wrong hose installed on equipment and pressure release
5. Poor hose location exposes failures to heat sources on the equipment
6. Overheated brakes ignite oil spray
7. Lights fault when oil sprays and fire starts.

In all, 14 unique, potential causes were identified and a BTA was performed on each. These 14 causes occupied the top event circle of the BTA (*Figure 8*). From the 14 BTAs, 34 existing prevention controls were identified (*Table 22*).

Table 22 – Existing key prevention controls for the underground workshop fire risk assessment.

TOP EVENT => Hydraulic oil leak/spray and ignites		
1. Worn hydraulic hose or coupler failure and pressure release	PC1	Hydraulic hose and couplings are designed to Standard (MH)
	PC2	Hoses are to be run away from hot areas or a barrier is located between hoses and hot spots (PB)
	PC3	The mine has a standard for 4 braided hoses and fittings that is currently being put in place (P)
	PC4	There is a specification on preventive maintenance program done on equipment that checks hoses and couplings (P)
	PC5	Inspections of equipment are done before work that looks for damaged or badly located or worn hoses (PST)
	PC6	When locomotive is shut down, the hydraulic pressure bleeds from the system (MH)
	PC7	Gage in the cab indicates pressure in hydraulics (WD)
2. Hot works damages hose	PC8	Procedures exist to do welding and cutting work in the workshop (P)
	PC9	Welding and cutting is not done on running (hydraulics are pressurized) equipment (P)
	PC10	In the rare case where welding and cutting works are done in proximity to hydraulic, the hydraulic lines are shielded (PB)
	PC11	Area of hot work and equipment is cooled by water hose after welding (P)
3. Unknowing removal of the hose under pressure (towed in or running)	PC12	Person should shut down and verify no hydraulic pressure before working on equipment (PST)
	PC13	Even if equipment is shut down, persons should check pressure gage before working on equipment (PST)
	PC14	Should be tag/note left on equipment if it has a problem (e.g. Needed to be towed in/pressurized) (WD)

	PC15	Completed work is recorded on Work Orders (P)
4. Hose pinched during installation or wrong hose installed on equipment and pressure release	PC16	Persons clean and inspect work after completion to identify issues such as pinched/damaged hoses (PST)
	PC17	Mine hose sizes are standardized (MH)
	PC18	Mine also gives standard specification for hose lengths to minimize the numbers of hose lengths (MH)
	PC19	Hoses are available in workshop or warehouse (P)
	PC20	There is a follow-up process to address the need for an incorrect (too long) hose with Work Order (P)
5. Poor hose location exposes failures to heat sources on equipment	As Above: See PC2 to PC5	
6. Overheated brakes ignite oil spray	PC21	If brakes are locked, then machine is not dragged to workshop (P)
	PC22	If partial brake fails (not locked), then machine is not dragged to workshop (P)
7. Lights fault when oil sprays and fire starts	PC23	Pit lights are designed to reduce hot surface/oil ignition exposure (MH)
TOP EVENT => Short circuit in battery cell leads to fire		
8. Dirty battery, battery fatigue, low water	PC24	Weekly inspection of equipment, check batteries and clean if required (P)
	PC25	Battery is also inspected by bottom attendants (P)
	PC26	Battery is also inspected in workshop before or after work and cleaned if required (P)
9. Battery age	PC27	Batteries are dated and replaced if date and cell status indicates (P)
10. Improper charging	PC28	Operators park and charge loco batteries in station (P)
	PC29	Battery attendees and mechanics also put locos at stations to charge (P)
	PC30	Battery should be fully charge and cooled for 8 hours (P)
TOP EVENT => Fire in garbage storage		
11. Spontaneous combustion in garbage storage	PC31	Garbage is changed out every 5 days at least (P)
	PC32	Hot items (>302 F) do not go into garbage (P)
	PC33	Items are put into garbage in bags (PB)
TOP EVENT => Exhaust DPM filter fire when withdraw from engine		
12. Hot materials enter DPM from catalyst, etc.	PC34	Reduce excess idling to lower hot materials entering DPM from catalyst (P)
TOP EVENT => Pool of oil from a failed pump ignites in the maintenance pit and leads to a fire		
13. Oil pump fails due to damage or seal failure		No current controls
14. Welding/cutting near pit ignites oil pool in pit		No current controls

PC – Prevention Controls
MH – Minimize Hazard
PB – Physical Barrier
WD – Warning Devices
P – Procedures
PST – Personnel Skills and Training

Based on the analysis that was done by the team, existing controls can be placed into one of the six categories described in Section 4.4. Twenty-seven percent of the key existing prevention controls (*Figure 15*) involve the design of equipment, i.e. minimize hazards (MH) and physical barriers (PB). For example, hydraulic hose and couplings are designed to standards with hoses located away from hot surfaces or barriers placed between hoses and hot spots. The mine has standards for four braided hoses and the size and lengths of hoses. In addition, pit lighting systems are designed to reduce hot surfaces and lower ignition potential. The diesel particulate modules are designed to reduce hot material risks.

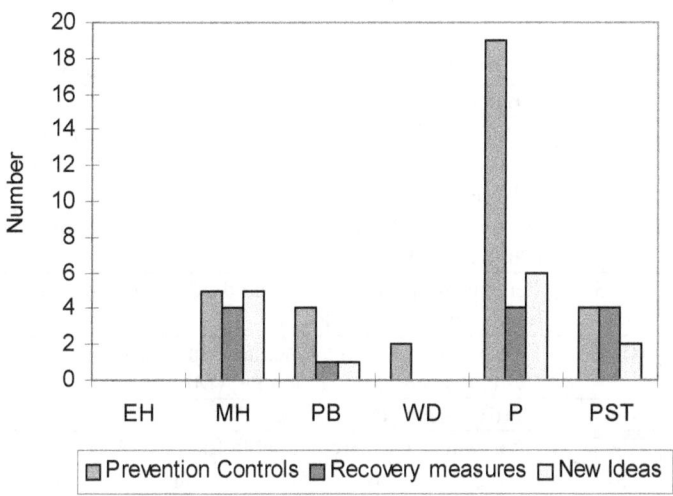

Figure 15 - Distribution of prevention controls and recovery measures for the underground workshop fire risk assessment.

Six percent of the existing prevention controls were classified as warning devices (WD). For example, each locomotive cab has a hydraulic pressure gauge to help identify that all pressure is bled from the system prior to maintenance activity (PM7). Also, equipment needing repairs is tagged and the condition of the hydraulic system (pressurized) is noted (PM14).

Sixty-eight percent of all existing prevention controls consist of procedures (P) and personnel skills and training (PST) that focused on <u>maintenance</u> and operational issues. Preventive maintenance checks are regularly done on hoses and couplings and on the location of hoses with respect to hot surfaces. Battery maintenance is also very important, with inspection and cleaning on a weekly basis. In addition, the shaft bottom attendant is required to inspect batteries on a regular basis. If equipment brakes are locked and cannot be released, the machine is hauled, not dragged, to the workshop. Each locomotive operator is required to inspect equipment prior to operation. They are trained to look for damaged, badly located or pinched hoses. They are also required to tag equipment if a problem is found. Maintenance personnel are required to shut down equipment and verify that no hydraulic pressure exists on the equipment prior to commencing work and to completing a work order. Batteries are replaced when their expiration date is reached. Locomotive operators park and charge batteries at designated stations and allow batteries to fully charge and cool for 8 hours.

Welding/cutting/grinding (hot works) and housekeeping are recognized as extremely important key controls for preventing workshop fires. SOPs exist for hot works in the workshop area. Hot work areas are cooled with water and hot works are never attempted on running or pressurized equipment. Housekeeping issues centered on placing oily rags in sealed bags prior to placing the bags in garbage storage areas. Garbage is changed out at least every 5 days and hot surfaces (>300° F) do not go into garbage.

Existing key recovery measures were also identified as part of the BTA. In the case of the underground workshop fire, the recovery measure was essentially the same for all 13 potential causes. Thirteen key recovery measures were identified for an underground workshop fire

(*Table 23*). The distribution of the control categories is shown in *Figure 15*.

Table 23 - Existing key recovery measures for the underground workshop fire risk assessment.

TOP EVENT => Fire in the workshop area, i.e., on the locomotive in garbage retainer, in the pit		
Fuel source comes in contact with heat source	RM1	Persons will try to shut off equipment in cab locations to stop hydraulic fluid spray (note: when equipment is in operation there must be an operator in the cab) (PST)
	RM2	Persons would used hand-held, readily accessible fire extinguishers to fight fire (20 lbs) (PST)
	RM3	Persons are trained to use hand-held fire equipment every year including hands-on exercises (PST)
	RM4	Fire suppression on a machine can be manually and automatically activated (will also shut down machine) (MH)
Small fire starts	RM5	Notify surface clerk of fire (P)
	RM6	Live water hoses are available for use (MH)
Fire persists or grows	RM7	Surface clerk gathers information and notifies state/federal regulatory agencies and shift management (minimum 5 min to take action) (P)
	RM8	Workshop has fire sprinklers activated by overhead fire sensors (MH)
	RM9	Mine has an underground and surface fire brigade (PST)
	RM10	Other fire fighting equipment is also located atop of shaft, as well as 8-10 headings away (P)
	RM11	There are W65 CO units and SCSRs on the mobile equipment and at designated locations near the workshop to aid in emergency egress (PB)
	RM12	Firefighters have PS5100 and PA100, etc., to fight large fire (MH)
Fire develops into major fire requiring emergency egress	RM13	A major fire in the mine requires mine evacuation and the shift foreman will decide on the evacuation route considering fire location (P)

RM – Recovery Measures
MH – Minimize Hazard
PB – Physical Barrier
WD – Warning Devices
P – Procedures
PST – Personnel Skills and Training

The key components to the existing recovery measures are fire fighting and emergency egress from the mine. Hand-held fire extinguishers (20 lbs) are readily accessible on all equipment and within the workshop area (MH). Miners are trained to use hand-held fire equipment every year including hands-on exercises (PST). Most mobile equipment repaired in the workshop has automatic fire suppression capability (MH). The workshop has been fitted with a sprinkler system, activated by overhead fire sensors. Live water hoses are available for use in the workshop area. Additional fire fighting equipment is located at the top of the nearby intake shaft, as well as 8-10 cross-cuts from the workshop (MH). Large fires can be fought with PS5100, PA100, etc., fire fighting equipment.

Workshop personnel are trained to immediately notify the surface clerk of fire (PST). The surface clerk gathers information, notifies shift management, and calls federal and state regulatory officials (minimum 5 min to take action). The mine has both an underground and surface fire brigade. If an emergency egress is necessary, miners are trained to don W65s and SCSRs and escape from the mine through primary or alternate escapeways (PST).

5.4.3.4 - Step 4, Identify New Prevention Controls and Recovery Measures

As part of the BTA, 14 new prevention control and recovery measure ideas were identified by the team during the risk assessment to further reduce the workshop fire risks at the mine (*Table 24*). Nine of the ideas are prevention controls and five are recovery measures.

Table 24 - New ideas for an underground workshop fire risk assessment.

New prevention control ideas	NI1	Reinforce the need to shut down equipment and inspect equipment prior to welding and cutting (P)
	NI2	Reinforce need to shield hydraulic systems near the source if hot work must be done on operating equipment (P)
	NI3	Remove all heat sources from the area of equipment to avoid ignition of any hydraulic pressure leaks (P)
	NI4	Investigate ways of being more systematic and thorough about equipment repair needs and status before or when equipment is taken to the workshop (P)
	NI5	Use the behavior controls in this risk assessment to focus the Safety Behavior Observations (SBO) program/activities in workshop (PST)
	NI6	Reinforce need to ensure that lights and covers in pit are in good condition (P)
	NI7	Add pit inspection to fire boss's job duties and consider inspection/monitoring by workshop personnel too. Make sure that these inspections examine for pooled oil in the pit. Add the requirement to inspect pit for pooled oil to pre-welding/cutting/grinding (any hot work) job preparation (P)
	NI8	Investigate modifying motors designed so they will not operate in low voltage (thereby they should be fully charged) (MH)
	NI9	Investigate changes that will allow battery removal and installing only fully charged batteries (MH)
New recovery measure ideas	NI10	Investigate adding a 150-lb wheeled dry chemical fire extinguisher (MH)
	NI11	Identify what the workshop personnel should do if hand-held fire extinguishers are not adequate (get other equipment or egress, breathing apparatuses, and include info in training) (PST)
	NI12	Decide whether doors in stoppings should be opened, allowing smoke into return (note that this would greatly increase airflow in workshop, note water line in return may be cut, note also idea to install water screen pipes in the return entry if doors are to be open) (PB)
	NI13	Consider the installation of a water source on the intake inside of the workshop (MH)
	NI14	Investigate getting a suitable foam system for pit to put out oil pool fire (MH)

NI – New Ideas
MH – Minimize Hazard
PB – Physical Barrier
WD – Warning Devices
P – Procedures
PST – Personnel Skills and Training

The 14 new ideas can be divided into design, operating procedures and fire/emergency response issues. Two design ideas were identified. The team recommended that management investigate 1) modifying motor designs so they will not operate at low-voltage (MH) and 2) modify chargers to prohibit battery removal without a full charge (MH). In both these situations, the low-voltage condition of the battery increases the likelihood of overheating which increases the potential for a fire. Both of these new ideas would limit the potential use of low-voltage batteries.

Numerous new ideas were presented that focus on operating procedures. Two ideas focus on reinforcing the need to shut down equipment and inspect equipment prior to maintenance work and to ensure that lights and covers in the pit are in good condition (PST). Prior to working on equipment, workers should remove all heat sources from the area to avoid ignition of any hydraulic pressure leaks. This could be accomplished with an SOP. The team also recommended that some kind of notification system be investigated to systematically identify the repair needs and their status before, or when, equipment is taken into the workshop (P). This could increase the workshop personnel's awareness of potential hazardous conditions. The team also suggested that the fire boss and workshop personnel inspect the pit regularly for oil and other flammable materials (P). This inspection should also be done by the workshop personnel prior to hot work activity. Lastly, the team recommended that a Safety Behavior Observation (SBO) program/activity be initiated for the workshop (PST). The mine operator has personnel skilled in this approach.

Several new ideas focused on <u>emergency response</u> issues. The team recommended that workshop personnel understand emergency response procedures and discuss specific fire scenarios during the training exercise. For example, workshop personnel should understand what impact opening the ventilation door in the return air stopping at the far end of the pit might have on a small fire (PB). This could be an issue since a water line is located in the return airway and might be useful in fighting a maintenance pit fire. Also, if hand-held extinguishers fail to put out a small fire, should workshop personnel obtain additional fire fighting equipment, don breathing apparatus, or evacuate the fire site and egress from the mine (PST)? The team also suggested that fire fighting capabilities (MH) in the maintenance pit be increased by adding one or more of the following items: 1) a 150-lb wheeled dry chemical fire extinguisher, 2) a portable foam generator, or 3) a water source on the intake side of the maintenance pit.

<u>5.4.3.5 – Step 5, Discuss Implementation, Monitoring and Auditing Issues</u>

The information provided in the risk assessment seemed accurate and the team functioned adequately, although the lack of participation from labor may have limited the team's composite knowledge of the fire hazards in the maintenance pit. However, the addition of outside fire experts helped to bolster the team's knowledge of fire prevention issues. The team had a strong focus on monitoring, administrative controls and training when identifying existing and new prevention controls and recovery measures (*Figure 15*). This will require the mine to vigilantly monitor and audit these controls. In this case study, the risk assessment did not recommend ways to eliminate the fire hazards from the workshop area. For example, was it possible to relocate the shop to an area that would effectively eliminate hazards? The exercise was successful in developing a range of new ideas that could produce quality barriers, controls and recovery measures that would further reduce the risk of a workshop fire.

5.5 – Water Inundation Risk Assessment Case Study

Mines Ea and Eb are operating near an abandoned mine and adits that present a threat of inundation (*Figure 16*). The abandoned mine and adits are in the same seam and their workings are partially flooded. The abandoned mine flooded area is large with an estimated water head of approximately 30 ft at Mine Ea and 80 ft at Mine Eb. The abandoned mine was extracted using the room-and-pillar method, creating a significant water reservoir. The adits are to the east of Mine Eb and have two parallel entries, several hundred feet long and connected by cross-cuts. These potential water reservoirs are relatively small and may contain dangerous gases.

All of these old mine workings are potentially filled with water and represent a significant hazard. Inundation hazards present risks for the mining operation. The mine has put into place many Best Practice controls to prevent an unwanted inundation event. The hazard is complicated by potential inaccuracies of existing abandoned mine maps and the lack of maps for the adits.

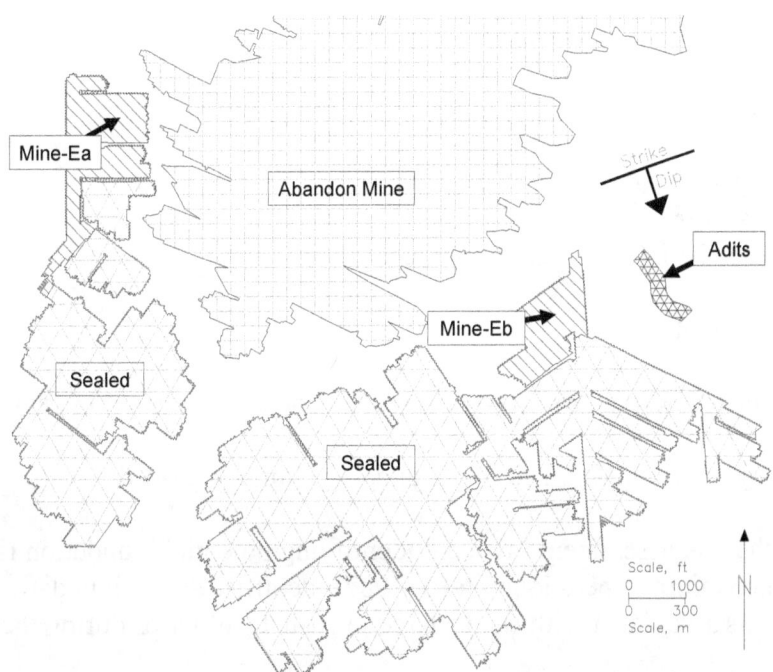

Figure 16 - Location of Mines Ea and Eb and adjacent water-filled abandoned mine and water/gas filled adits.

The two mines are underground room-and-pillar coal mines with entries less than 48 inches high by 18 feet wide. These mines have been able to maintain a modest size footprint by periodically sealing mining sections as activity in the area ceases. Each mine employs approximately 35 miners and operates a single auger type continuous mining machine. Current working faces are mostly located down-dip from the abandoned mine. The operator attended the NIOSH-sponsored workshop on Minerals Industry Risk Management and identified the mine inundation hazard as the most significant risk.

5.5.1 - Risk Assessment Scope

The objective of this risk assessment was to 1) identify hazards associated with the potential for an inundation at Mines Ea and Eb, 2) evaluate strategies and techniques to lessen the risk associated with an inundation, and 3) develop an action plan for new ideas. The mine operator was also interested in developing an inundation management plan. The mine has examined the possibility of pumping out all water from the abandoned mine but deemed it to be impractical due to the size of the water body. Elimination of the hazard was therefore not considered by the risk assessment team. Numerous controls, both required and in addition to MSHA regulations, are currently used at both mines. The mine was interested in examining additional controls to further lower the risk of inundation.

5.5.2 - Risk Assessment Team

The team was made up of persons employed at Mines Ea and Eb, as well as from the parent company and a NIOSH representative, as follows:
    Two management representatives
    Two shift foremen
    One Miner
    One Engineer
    One Geologist
    NIOSH observer
    Facilitator – MISHC (University of Queensland)

5.5.3 – Risk Assessment

All five steps of the MHRA approach were followed and are discussed in detail below.

5.5.3.1 – Step 1, Identify and Characterize Inundation Mining Hazards

The first step in the risk assessment involved identifying potential inundation issues around the current and planned mining operations. Jobs (1987) identified seven inundation sources and the number of accidents associated with each source in British collieries during the period 1851 to 1970:
1. Contact with surface water – pond, river, canal or stream (9)
2. Contact with surface unconsolidated deposits – glacial or organic (8)
3. Strata water entering the mine workings (2)
4. Shaft sinking (4)
5. Clearing old shafts (14)
6. Contact with abandoned old workings (162), and
7. Failure of an underground dam, seal or leakage of a borehole (9).

Significantly, 78% of the accidents were associated with contact with abandoned old workings. The team discussed the different sources of inundation and determined four potential hazards existing at Mines Ea and Eb:

1. Water from the up-dip abandoned mine,
2. Water and gases from the adjacent adits,
3. Water from surface creeks, and
4. Water from surface drainage.

Water from the abandoned mine is a potential hazard, with approximately 30 to 80 ft of water head. These flooded workings are in the same seam as Mines Ea and Eb and are generally at the same or higher elevations. There are also small adits open to the surface along the coal seam outcrop containing either water and/or gases typically found in old mine openings. These adits generally consist of two headings connected by cross-cuts and driven from the coalbed outcrop a few hundred feet into the hillside. The maximum adit water head is estimated to be 5 ft. There are no plans or maps in existence to help locate these entries in relation to Mine Eb.

Surface creeks run over Mine Ea. The streams have relatively low volumes of water and no water from the surface has been detected in the mine. There is also intermittent drainage from surface water. Recently a 100-year rainfall event entered the box-cut of Mine Eb and flooded its working faces. Subsequently the box-cut design was changed to reduce the 100-year rain event impact. The risk assessment team considered the surface creeks and surface drainage hazards to be minor and decided to focus on the water hazards from the abandoned mine and adjacent adits.

The risk assessment geographic boundaries are defined by the outline of the current and future face developments adjacent to the abandoned mine and adits (*Figure 17*). Three segments define these boundaries:
    Segment 1 represents Mine Ea's current and future face developments close to the projected boundary of the abandoned mine,
    Segment 2 is for Mine Eb's current and future face developments close to the boundary of the abandoned mine,
    Segment 3 is representative of the boundary between Mine Eb's future face developments and the potential location of adits.
These three segments are the geographic focus of the risk assessment team.

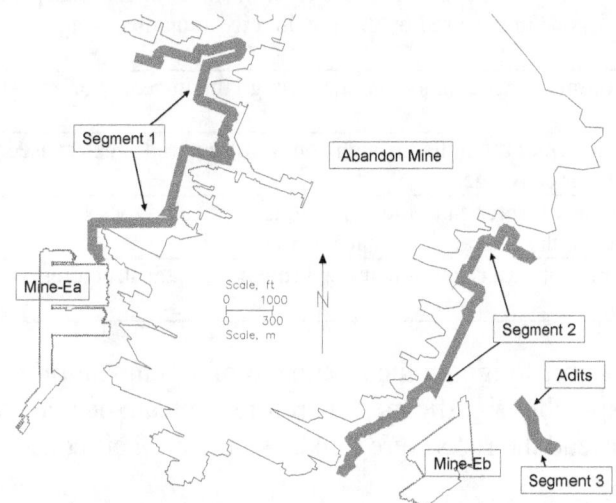

Figure 17 - Location of geographic boundaries of the risk assessment. Thick lines define the boundaries between the abandoned mines and the current projections for Mines Ea and Eb.

5.5.3.2 – Step 2, Rank Potential Unwanted Events

With step 1 complete, the team determined six possible mechanisms for water to enter the active mines from the abandoned mine and adits (*Table 25*). These possible mechanics were all examined for their ability to produce an inundation event with high consequence.

- Two of the mechanisms are associated with the active mines penetrating the abandoned mine and adits. The negative consequence of these events would be very high.
- A third mechanism saw in-seam horizontal drilling activity as a means of creating a possible avenue for water and gas flow. The consequence of this potential unwanted event is thought to be moderate since the drill hole is relatively small and the drills are fitted with packers and shut-off valves capable of isolating the borehole water from the mine.
- A fourth mechanism has moderate volumes of water entering the active mines through isolated areas of excessive roof-to-floor convergence. These areas are capable of locally increasing water flow rates. Here again the consequence is considered to be moderate since the water volumes are not expected to be significant.
- The final two mechanisms had relatively small quantities of water flowing along geologic structures within or adjacent to the mined coalbed. A study by Moebs and Sames (1989) characterized the water flows rates along a potential geologic discontinuity that produced a connection between a water-filled abandoned mine and an active mine workings. These water flow rates represented a concern for the mine operator, but were unlikely to result in a catastrophic release of water. Therefore, the consequence of these events is considered to be low.

Table 25 - Consequences of different inundation mechanisms.

#	Mechanism for inundation	Consequence
1	Water violently enters the mine under pressure through a relatively large opening caused by mining directly into the abandoned mine along Segment 1	High
2	Water violently enters the mine under pressure through a relatively large opening caused by mining directly into the abandoned mine along Segment 2	High
3	Considerable quantity of water and/or mine gases enter the mine under relatively low pressure through an opening caused by mining directly into adits along Segment 3	High
4	High-pressure, low-volume water enters the mine through the in-seam horizontal drill holes	Moderate
5	Moderate water volumes enter the mine through zones of fractured rocks caused by excessive roof-to-floor convergence	Low
6	Relatively small volumes of water enter the mine along permeable rock layers and geologic structures in the coal and its adjacent strata	Low
7	Relatively small volumes of water enter the mine through the cleat structures within the coal seam	Low

In this study, it was possible to ignore the likelihood of an inundation event because the consequences are indisputably significant. Therefore the team decided to forego ranking the risk using a risk matrix. Instead the risks were ranked solely on their consequence.

5.5.3.3 – Step 3, Determine Important Existing Prevention Controls and Recovery Measures

A BTA was used to determine important existing and new prevention controls and recovery measures. The high-consequence events listed in *Table 25* were combined to read "water inundation occurs from adjacent old mine workings" and placed within the central node of the BTA (*Figure 18*). Three threats and three consequences were identified by the team and discussions of existing and new prevention controls and recovery measures followed. The threats were: 1) didn't know old mine workings were there; 2) dangerous gases in the adits; and 3) mined into old workings due to mining error. The consequence of mining into an old mine workings were: 1) increased water flow into active mining area; 2) minor water inrush; and 3) major inrush blocks normal egress routes.

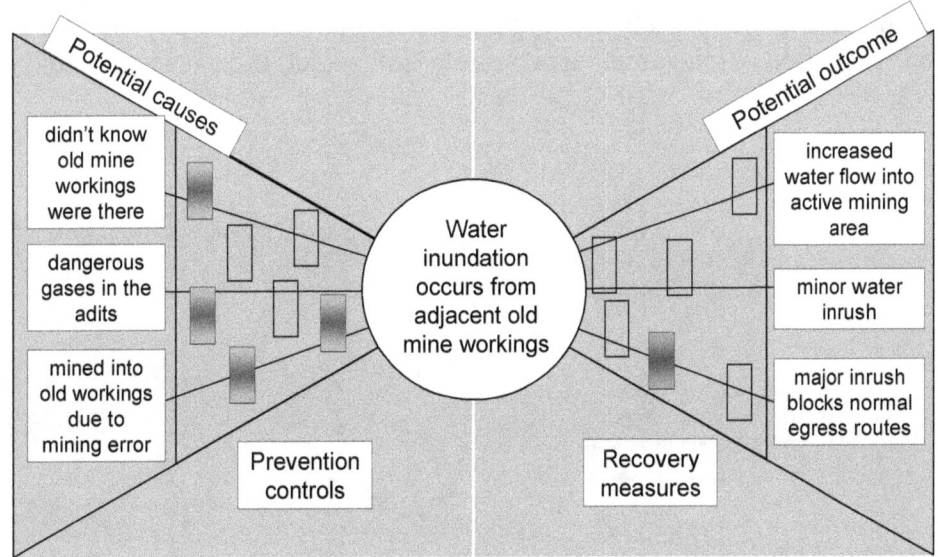

Figure 18 - Graphical depiction of the BTA used in the inundation risk assessment.

The mine operator is using several Best Practice techniques to aid in preventing the unwanted event. The techniques are listed below and are used in many different scenarios, principally to help verify the location of old mine workings. These techniques were often used as priority prevention controls during discussions of specific threats and consequences. For the purposes of this study, the four techniques were combined into one prevention control aimed at verifying the location of old mine workings.

- Technique 1 - Discussions with miners familiar with the abandoned mine in question: The abandoned mine was in operation until 1967. There are miners available who have knowledge of this recently abandoned mine plan and are used to help verify the accuracy of existing mine maps. On the other hand, adits were developed long ago and no records or miners exist with knowledge of these mines.
- Technique 2 - Surface seismic reflections capable of detecting underground mine voids: Several surface seismic reflection lines have been run over areas that were relatively far from the active mine but where additional information about the location of the abandoned mine was desired (*Figure 19*).

- Technique 3 - Surface geophysical resistivity surveys capable of detecting water-filled mine entries: Resistivity surveys have been completed, probing for water-filled entries along the boundaries of the abandoned mine (*Figure 19*).
- Technique 4 - Surveyed boreholes both from the surface (vertical) and underground (horizontal) to pinpoint the location of old mine workings: Boreholes are used to confirm the location of old mine workings. Surface boreholes have helped to establish the elevation of the water pool in the abandoned mine and to find adits. These holes are capable of removing water from the abandoned mine and present a safe way to ventilate gases from the adits. Horizontal drill holes are a much more effective means of locating old works but require the underground mine to be in proximity to the old mine workings. Both mines have used this technology (*Figure 19*). Typically, the boreholes are drilled from the inby end of a development section. These boreholes either probe the coalbed ahead of the working faces or examine the coalbed to be mined by future development sections. Typically, horizontal boreholes are used if further clarification is required or when mining occurs closer than 200 ft from a known old mine workings.

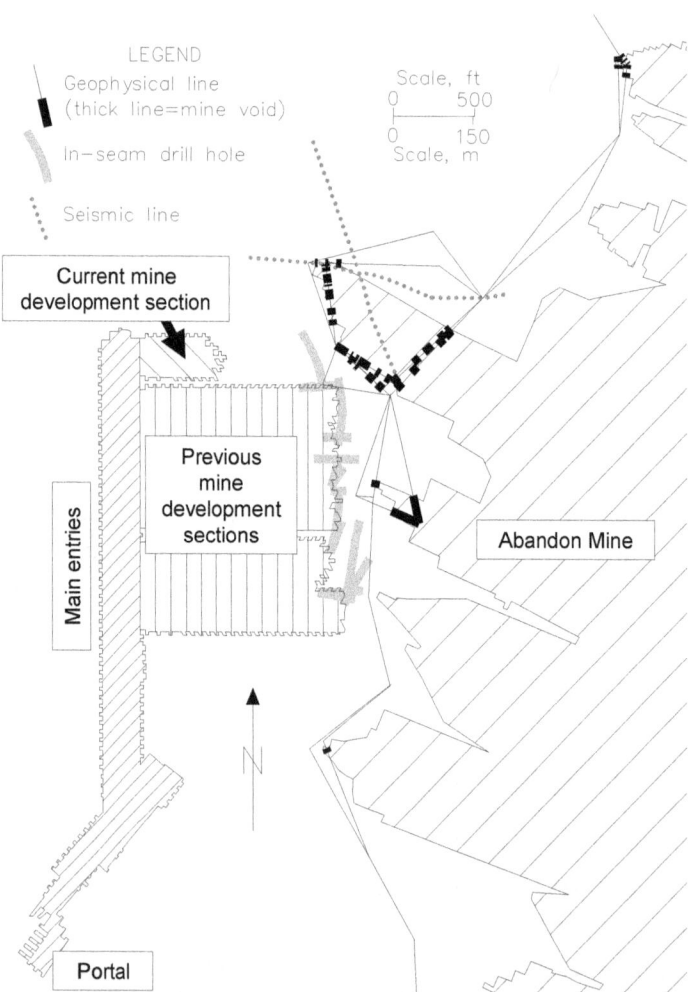

Figure 19 - Techniques used to find the location of water-filled old mine workings at Mine Ea.

Besides the four techniques to verify the location of old mine workings (PC1), the team identified six other existing prevention controls (*Table 26*). Five controls were aimed at lessening the risk of mining into old workings because their locations were not known. These controls focused on verifying the true location of the old mine workings (PC1), identifying a 200-ft barrier around these old mine workings to account for inaccurate data or interpretations (PC2), in-seam probe drilling when faces were thought to be less than 200 ft away (PC3), ensuring communications to all necessary personnel (PC4), and observing unusual conditions that might signal that old mine workings were close by (PC5). The sixth control was specific to adits, where vertical boreholes are sometimes used to ventilate potentially dangerous gases along Segment 3 (PC6). The seventh control is used to lessen the chances of a mining error occurring by requiring mining crews to perform daily survey checks and compare them to mine projections (PC7).

Table 26 - Summary of existing prevention controls and recovery measures for a potential mine inundation.

Existing prevention controls		
Threat 1 – Didn't know old mine workings were there	PC1	Use the four identified techniques to assist in validating the location of old mine workings (MH)
	PC2	Identify a 200-ft barrier of solid coal around the known position of old mine workings, place its position on mine maps, and communicate this information to the workforce (PB)
	PC3	In-seam horizontal drilling is required if the active mining faces are within 200 ft of the known location of old mine workings (MH)
	PC4	Communicate the information about efforts to validate the location of old mine workings and their known locations to miners (PST)
	PC5	Workers observe certain conditions that might indicate that a water-filled old mine workings is close, i.e. enhanced water making its way through the coal seam at the working face or the smell that standing water sometimes gives off, and communicate these conditions to their foremen and others (PST)
Threat 2 – Dangerous gases in the adits	PC6	Vertical boreholes are used to ventilate potentially dangerous gases from adits in Segment 3 (MH)
Threat 3 – Mined into old workings due to error	PC7	Mining crews are required to perform daily survey checks and compare to mine projections (P)
Existing recovery measures		
Consequence 1 – Increased water flow into active mining area	RM1	Operator will call others underground on hand-held radios to alert them of changing conditions and give evacuation instructions (P)
Consequence 2 – minor inrush	RM2	Follow existing Emergency Response Plan to evacuate the mine (PST)
Consequence 3 – major inrush blocks normal egress routes	RM3	Follow existing Emergency Response Plan to evacuate the mine; however, exception could occur, i.e. miners in Mine Ea may decide to move to the highest elevation, potentially north along the mains, to escape inrush as it moves from the faces down-dip to the portal; and miners in Mine Eb may need to alter egress routes as they escape up-dip toward the portals and away from the flooding faces. (PST)

PC – Prevention Controls
RM – Recovery Measures
MH – Minimize Hazard
PB – Physical Barrier
WD – Warning Devices
P – Procedures
PST – Personnel Skills and Training

Three existing recovery measures were also identified. If mining conditions noticeably change or an inundation occurs, workers communicate information and evacuation instructions through hand-held radios (RM1). When water inrush conditions occur, the miners are to evacuate according to the Emergency Response Plan (RM2). Finally, if a major inrush were to occur, it could possibly block normal egress routes. Miners may consider alternate egress routes as outlined in RM3.

5.5.3.4 – Step 4, Identify New Prevention Controls and Recovery Measures

Five new ideas were identified by the team to further reduce the inundation risks at the mine (*Table 27*). These ideas were listed as part of an Action Plan with the ideas inserted and space left for derivation of specific actions, timing and resources. Four of the five were potential recovery measures. The only new prevention control formalizes the process of using water/smell conditions to look for or to identify potential inundation conditions (NI1). The team thought it important to have calculations of potential flooding rates for both mines (NI2). This information could be used to further evaluate the inundation hazards and could be used when considering NI4 and 5. The team believed that an important way to minimize losses was to restrict access to certain areas of the mine during key time intervals (NI3). Restricted access should be considered when mining in areas near the abandoned mine and adits, i.e. along Segments 1, 2 and 3. The team spent considerable time discussing how egress from the faces would be influenced by the size and location of the inrush event. It was not possible to sufficiently analyze this issue during the risk assessment and make adjustments to the Emergency Response Plan (NI4). For example, it may not be possible for the miners to use existing escapeways if the inundation event occurred along Segment 1. In this case, water would enter the face and run down-dip along the main entries, forcing miners up-dip toward the working faces along the main entry. Other scenarios were also discussed with some having the miners wait in less hazardous areas until the threat subsides or they are rescued. The team thought it important that techniques be investigated (NI5) to communicate, locate and rescue trapped miners underground (e.g. tapping, signals, phones, surface access, emergency supply skid, etc.).

Table 27 - New ideas proposed by the risk assessment team for preventing or recovery from an inundation at Mines Ea and Eb.

New prevention control ideas	NI1	Formalize water/smell conditions to look for or to identify possible inundation, as well as action to be taken in those conditions, and introduce to personnel (P)
New recovery measure ideas	NI2	Calculate rates of flooding for both mines to estimate the consequences of an inundation on egress routes and communicate to the miners (PST)
	NI3	Restrict access to certain areas of the mine, i.e., no one in return entry when mining in areas near the abandoned mine and adits. (Segments 1, 2 and 3) (P)
	NI4	Develop the Emergency Response Plan beyond MSHA requirements to address specific mine issues (P)
	NI5	Investigate methods and actions that should be undertaken if persons are trapped underground due to flooding (e.g. tapping, signals, phones, surface access, emergency supply skid, etc.) (PST)

NI – New Ideas
MH – Minimize Hazard
PB – Physical Barrier
WD – Warning Devices
P – Procedures
PST – Personnel Skills and Training

5.5.3.5 – Step 5, Discuss Implementation, Monitoring and Auditing Issues

Assuming that the information provided in the risk assessment was accurate, completion of the Action Plan and an increased focus on monitoring and auditing of the key identified controls would appear to provide an opportunity to effectively reduce the risk of fatalities related to inundation at Mines Ea and Eb. The risk assessment showed that the mine relied extensively on prevention controls (PC) that reduce the risk of water inundation from mining into the abandoned mine (*Figure 20*). The figure also shows a somewhat limited number of existing recovery measures (RM) identified by the risk assessment team. Four of the five new ideas addressed recovery measures, demonstrating the team's interest in improving the way the mining operation should respond to an actual inundation event. All of the new ideas were controls classified as procedures (P) and personnel skills and training (PST) (*Figure 20*).

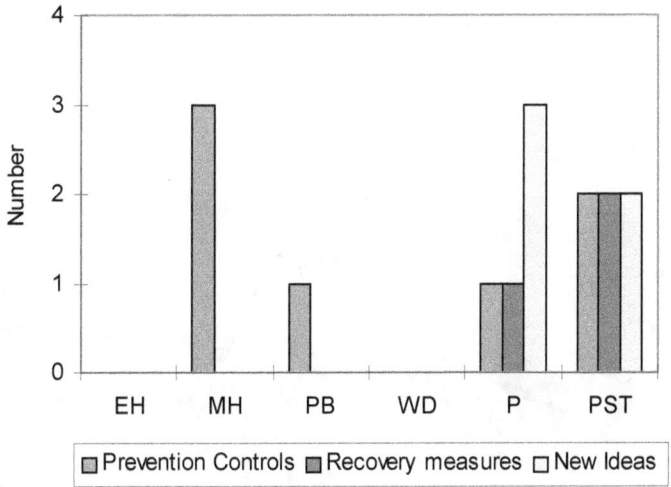

Figure 20 - Distribution of prevention controls and recovery measures for the water inundation risk assessment.

5.6 – Escapeway Egress Blockage Risk Assessment Case Study

Mine F is an underground limestone mine experiencing unstable ground conditions that potentially threaten the use of its alternate escapeways. The general conditions found at the mine are shown in *Figure 21*. The part of the mine relevant to this study was mined over 40 years ago using the room-and-pillar technique. Large rooms were driven 45 ft wide and 30 ft high perpendicular to a highwall in an adjacent quarry and off-set cross-cuts of the same size were mined typically on 90-ft centers. The parallel Primary and Alternate Escapeways run southeast from Portals No.1 and 2 to inby portions of the mine. In January 1994, a roof collapse occurred in an area adjacent to the Alternate Escapeway about 250 ft from Portal No.2. Between January 1994 and December 2006, other roof falls have occurred to the southwest of the Alternate Escapeway, resulting in a large restricted area. Management has responded to this roof instability hazard through Best Practice controls, including roof monitoring, supplemental standing support, and tensioned cable bolts in the Alternate Escapeway adjacent to the restricted area.

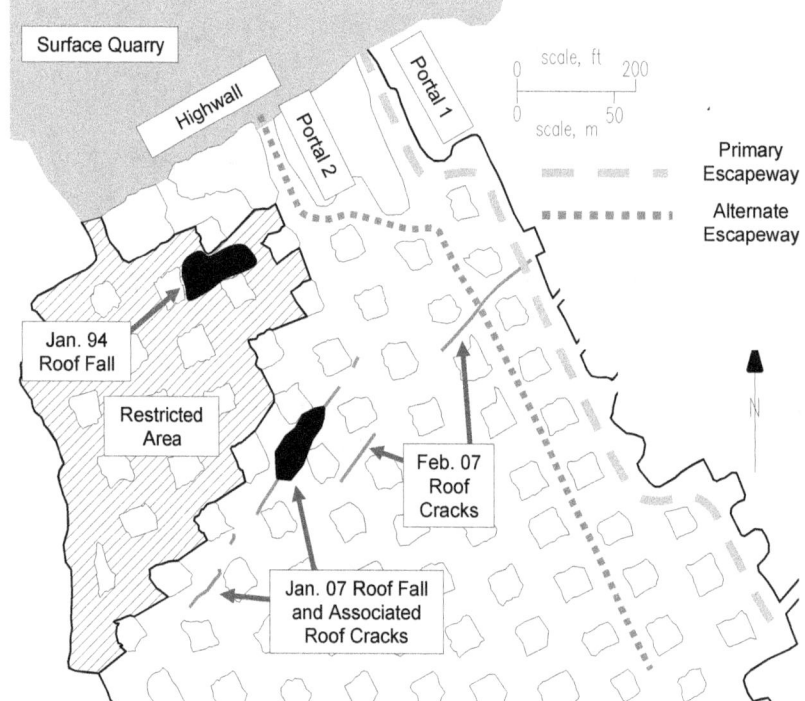

Figure 21 - Escapeways, roof falls and recent roof cracks found at the mine.

Recently, roof conditions in the escapeways showed signs of deterioration. Of particular concern are the January, 2007, roof fall and the January/February, 2007, appearance of intermittent, en echelon roof cracks (*Figure 21*). One of these roof cracks is especially troublesome because it extends across the Alternate Escapeway and into the Primary Escapeway, signaling an elevated risk.

5.6.1 – Risk Assessment Scope

The objective of this risk assessment is to 1) identify hazards that could affect egress through the mine's escapeways, 2) determine what unwanted events pose the greatest threat to mine workers escaping from the mine, 3) review the existing prevention controls and recovery measures, and 4) recommend new ideas to prevent, or recover from, potential disruption of escapeway egress. The initial MHRA steps consist of scoping document generation, scoping team selection, and assessment framework identification.

The risk assessment team agreed to frame the assessment by limiting it to the Primary and Alternate Escapeways when egress was disrupted by a roof collapse or fire hazard. Normal ventilation operating conditions were considered, which means the fan at the ventilation shaft is either exhausting or blowing into the mine. During exhaust conditions both escapeways are in fresh air, while under blowing conditions the escapeways will be in return air. Hazards and risks were considered in relation to their probability of occurring within five years.

5.6.2 - Risk Assessment Team

The scoping team consisted of the following persons:
- Mine Supervisor
- Mine Engineer
- Rock Mechanics Engineer
- Miner
- Safety Officer
- Subject matter experts (Strata Control, Ventilation, Mining Regulation and Mine Evacuation)
- Facilitator – NIOSH

5.6.3 – Risk Assessment

First, major hazards associated with egress through Primary and Alternate Escapeways during an emergency at the mine were reviewed. Consequences associated with the unwanted event were investigated and the likelihood of the event occurring was estimated. Threats that disrupt egress through the escapeway were analyzed and ranked using a risk matrix technique. Finally, existing and new controls and recovery measures were identified.

5.6.3.1 – Step 1, Identify and Characterize Major Potential Mining Hazards

Hazards that affect egress through the mine's escapeways are identified by first dividing the escapeway system into logical segments and then analyzing the various types of hazards. For the purposes of this analysis, the description of a metal/nonmetal escapeway follows from the definitions cited in the Code of Federal Regulations, Part 30, Section 57.11050 (CFR, 2005). The escapeway system at the study mine can be subdivided into six segments (*Figure 22*).

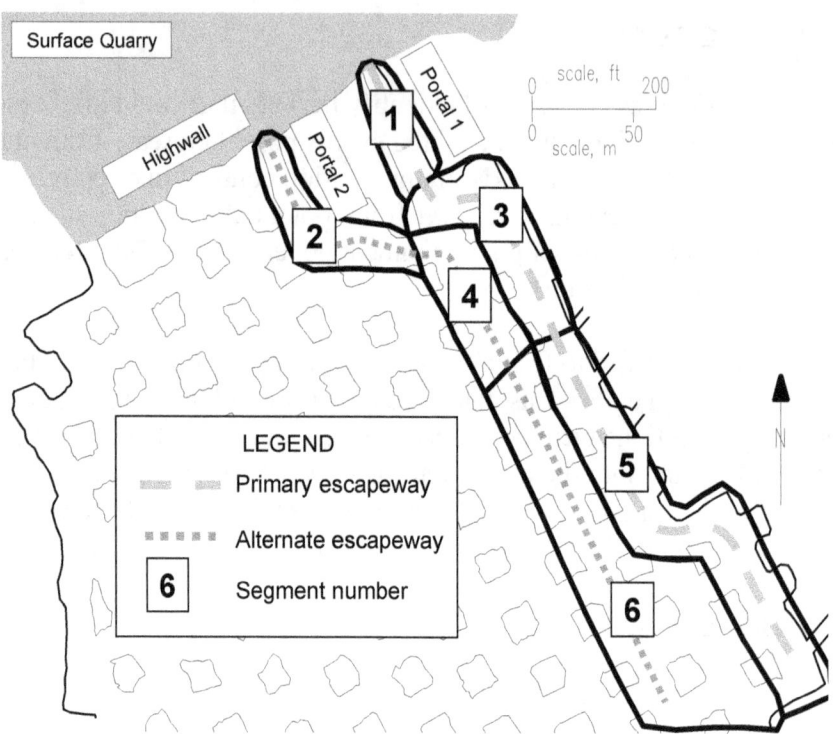

Figure 22 - Six segments of the mine's escapeway system.

The two kinds of hazards investigated in this study are fire and roof collapse. Fire hazards are identified by considering potential fuel and ignition sources. The results are summarized in *Table 28*.

Table 28 – Fire hazards consisting of potential fuel and ignition sources.

Fuels	Diesel equipment – truck, front end loader, backhoe, grader, crane, scoops and other smaller pieces of diesel equipment
	Fuel Storage – diesel tanks and other flammable materials
	Electrical – Mine carts, transformers, substations and power lines
	Other Equipment and Storage – conveyor belt, natural gas pipe line, wood, PVC pipe, and other minor amounts of material
Ignition/ heat sources	Overheating of diesel equipment, electrical equipment and electrical cabling
	Welding and cutting operations
	Lightning

Roof instability hazards are considered only in terms of their potential to block egress through escapeways. Small roof falls that can result in injuries were therefore excluded from the analysis, since they do not block egress. A NIOSH-developed tool, called the Roof Fall Risk Index (RFRI), was used to systematically identify roof fall hazards in the escapeways. The RFRI is a hazard assessment technique that maps the spatial distribution of stability conditions. The RFRI focuses on the character and intensity of defects associated with specific roof conditions (Iannacchione et al., 2006; Iannacchione et al., 2007). Ideally, values approaching 0 represent safer roof conditions, while an RFRI approaching 100 represents a serious roof fall hazard. The RFRI values for the mine's escapeway system are shown in *Figure 23*. Higher values indicate increasing risk of roof collapse in the absence of additional roof stabilization efforts. For example, the relative roof fall risk in Segment 1 of the Primary Escapeway is potentially lower than in Section 2 because this section contains roof bolts, wire mesh and narrower entry spans.

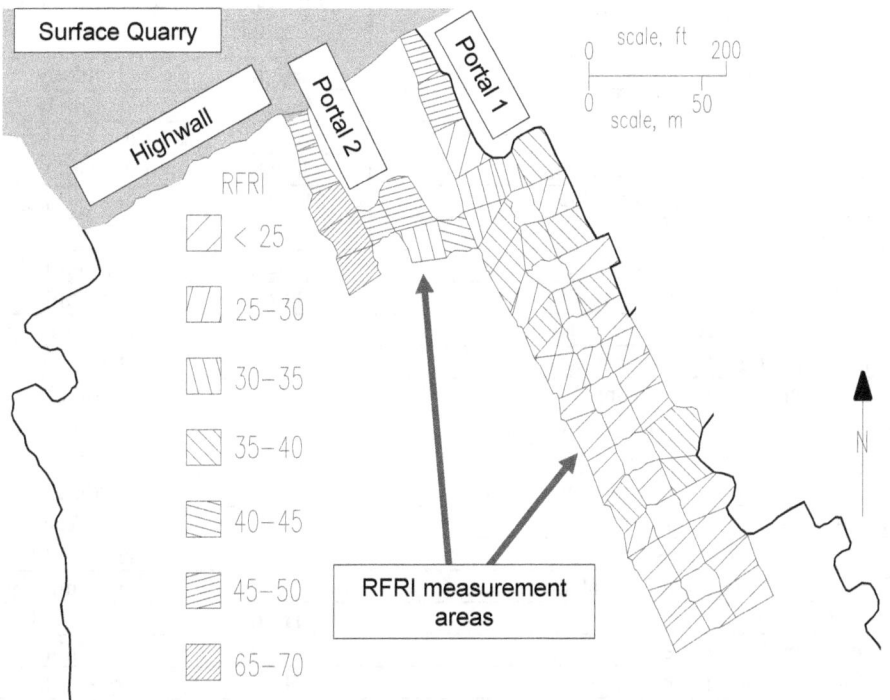

Figure 23 - Roof Fall Risk Index (RFRI) measured in the mine's escapeways.

5.6.3.2 – Step 2, Rank the Potential Unwanted Events

As the team became familiar with the escapeway routes, current ground conditions, and ventilation and operational requirements, the risk for a potential unwanted event in each segment was determined. The risks associated with unwanted events were rated using the WRAC method which considers the likelihood and consequences of each event.

The scoping team identified 28 potential threats based on the defined list of hazards (*Table 29*). Each potential threat was risk ranked using a qualitative risk analysis method and a 4 by 5 risk matrix (*Table 30*). Lower numbers indicate a higher risk. The likelihood of an event was subjectively assessed by considering the probability of the event occurring in the next five years. The consequences of an event were assessed by considering its potential impact on the ability to evacuate the mine in case of an emergency. This included consideration of blockage of escapeway routes and the spread of toxic fumes or smoke. Both exhaust and blowing ventilation scenarios were considered. The inability to use either escapeway for egress from the mine during an emergency was considered to be the highest impact consequence.

Table 29 - Risk ranking of potential threats grouped by escapeway segment.

Escapeway Segment	Potential Threats	Consequence (impact)	Likelihood (next 5 years)	Risk ranking
1	Equipment fire – fan exhausting	High	Unlikely	7
	Equipment fire – fan blowing	Moderate	Unlikely	8
	Roof collapse	High	Very Unlikely	11
	Diesel storage fire – fan exhausting	High	Very Unlikely	11
	Diesel storage fire – fan blowing	Moderate	Very Unlikely	16
2	Roof collapse	High	Likely	4
	Equipment fire - fan exhausting	High	Very Unlikely	11
	Equipment fire – fan blowing	High	Very Unlikely	11
	Electrical cable fire	Low	Very Unlikely	18
3	Equipment fire – fan exhausting	High	Unlikely	7
	Equipment fire – fan blowing	Moderate	Unlikely	12
	Charging station fire – fan exhausting	High	Likely	4
	Charging station fire – fan blowing	Moderate	Likely	8
	Transformer fire – fan exhausting	Low	Unlikely	16
	Transformer fire – fan blowing	Low	Unlikely	16
	Natural gas leak explosion	High	Very Unlikely	11
	Flammable storage cabinet catches fire	Low	Very Unlikely	20
	Roof collapse	High	Very Likely	2
4	Equipment fire – fan exhausting	High	Very Unlikely	11
	Equipment fire – fan blowing	High	Very Unlikely	11
	Roof collapse	Moderate	Very Likely	5
5	Equipment fire – fan exhausting	High	Unlikely	7
	Equipment fire – fan blowing	Moderate	Unlikely	12
	Roof collapse	Moderate	Unlikely	12
	Transformer catches fire	Low	Very Unlikely	20
6	Equipment fire during travel – fan exhausting	Low	Very Unlikely	18
	Equipment fire during travel – fan blowing	Low	Very Unlikely	18
	Roof collapse	Low	Unlikely	16

Table 30 - A 4 by 5 risk matrix for ranking the potential threats.

Consequence	Likelihood (event occurs in next 5 years)				
	Certain	Very Likely	Likely	Unlikely	Very Unlikely
High Impact	1	2	4	7	11
Moderate impact	3	5	8	12	16
Low impact	6	9	13	17	20
No impact	10	14	18	21	23

The top four potential threats identified through the WRAC are:
1. Roof collapse in Primary Escapeway of Segment 3
2. Charging station fire in Primary Escapeway of Segment 3
3. Roof collapse in Alternate Escapeway of Segment 2
4. Roof collapse in Alternate Escapeway of Segment 4

5.6.3.3 – Step 3, Determine Important Existing Prevention Controls and Recovery Measures

The team discussed the nature and quality of the prevention controls as part of the BTA. The outcomes of the BTA are presented in Appendix B. Eleven key prevention controls currently in place are identified and listed in *Table 31*. Controls for roof collapse hazard used in portions of the Primary and Alternate Escapeways consisted of 6- and 8-ft grouted bolts placed on 5-ft centers (PC1). This support occurs through most, but not all, of the Primary Escapeway and only within the first 200 ft of Portal 2 in the Alternate Escapeway. Typically the mine is completely scaled every 3 to 6 months, but scaling of individual loose rock occurs as needed (PC2). Periodic observations of roof conditions are made by the mine supervisor, mine engineer and miners on a regular basis (PC3). Both multipoint roof sag extensometers and roof-to-floor convergence sensors have been used in the Alternate Escapeway to assess stability conditions (PC5). Over 100, 30-ft long, 60-ton capacity cable bolts have been installed and tensioned to 20 tons as a means of adding support to a section of Alternate Escapeway roof near the restricted area (PC6). These cables were installed at a 30° angle from vertical to help prevent roof instabilities associated with prominent joints. Finally, three massive breaker wall standing supports, 45 ft wide by 30 ft high, were installed in cross-cuts between the Alternate Escapeway and the restricted area (PC7). Fire hazard controls consist of weekly battery checks (PC4).

Table 31 - Existing prevention controls and recovery measures for a loss of emergency escapeway at Mine F.

Existing prevention controls		
Threat 1 – Roof collapse in Primary Escapeway Segment 3	PC1	Primary support – 6- and 8-ft grouted bolts (PB)
	PC2	Scale roof and ribs every 3 to 6 months, or as needed (P)
	PC3	Periodic observation of roof conditions (PST)
Threat 2 – Charging station fire in Primary Escapeway Segment 3	PC4	Battery water levels and terminals are checked weekly (P)
Threat 3: Roof collapse in Alternate Escapeway Segment 2	PC5	Monitoring with multipoint extensometers (WD)
	PC6	Cable bolt support with steel screen (PB)
	PC7	Breaker wall standing support (PB)
	As above: see PC3	
Threat 4 – Roof collapse in Alternate Escapeway Segment 4	PC8	Monitor microseismic emissions from the mine (just begun) WD)
	As above: see PC2 & PC3	
Existing recovery measures		
Consequence 1 – Roof collapse blocks the Primary Escapeway	RM1	Escapeway must be cleared and re-supported or a new escapeway designated (PB)
Consequence 2 – Charging Station Fire in Primary Escapeway	RM2	Station partially enclosed by a block wall & metal roof (PB)
	RM3	Scoop has fire suppression system (MH)
	RM4	Fire extinguishers are present, although current policy is to evacuate rather than fight fire (PST)
	RM5	Main office coordinates communication via radio (P)
	RM6	Use radios to communicate fire alarm to all underground (WD)
	RM7	Sound the siren (WD)
	RM8	Lifelines exist in part of the Primary Escapeway (PST)
Consequences 3 and 4 - Roof fall blocks Alternate Escapeway	As above: see RM1	

PC – Prevention Controls
RM – Recovery Measures
MH – Minimize Hazard
PB – Physical Barrier
WD – Warning Devices
P – Procedures
PST – Personnel Skills and Training

The team identified nine existing recovery measures. The only recovery measure for a large roof collapse capable of blocking the Primary or Alternate Escapeways is to clean up the fall material and re-support the entry or develop a new Primary or Alternate Escapeway (RM1). Seven recovery measures for a fire in the charging station area were identified. A cinder block wall and metal roof have been built and could partially contain a charging station fire (RM2). Several pieces of diesel equipment have fire suppression systems (RM3). Fire extinguishers are present in this area, although current policy is to evacuate rather than fight the fire (RM4). In the event of an evacuation, radio communication, directed by the mine office, would be used to communicate a fire alarm (RM5). All miners and persons accompanying visitors are issued radios that can be used to communicate a fire alarm (RM6). A site-wide siren is also available at the main office that can be heard by all personnel outside the mine (RM7). Lastly, lifelines exist in part of the Primary Escapeway (RM8).

5.6.3.4 – Step 4, Identify New Prevention Controls and Recovery Measures

Fifteen new ideas were identified by the team (*Table 32*). New prevention controls are aimed at either the roof collapse or fire hazard. For the roof collapse hazards, new controls are divided into three groups: administrative, monitoring and engineering. An administrative control in the form of a policy could restrict personnel access to the Alternate Escapeway except in an emergency situation (NI1). A number of monitoring controls were discussed including: a regularly scheduled visual observation plan of roof conditions (NI2); installation of additional roof and crack monitors (NI3); repair and replacement of 12-year-old multipoint sag extensometer (NI4); and a trigger action response plan (TARP) for monitors (NI5). New ideas for preventing roof collapse hazards include a supplemental rock reinforcement program for Segment 3 of the Primary Escapeway (NI6). This should be done after additional monitoring has gathered sufficient information. A design should also be considered for stabilizing Segment 4 of the Alternate Escapeway (NI7). The purpose of this design would be to protect the growth of roof falls in the restricted area from weakening the roof in the Alternate Escapeway. The new control for the fire hazard was to place the charging station outside the mine (NI8).

Table 32 - New ideas proposed for preventing or recovery from a loss of emergency escapeway at Mine F.

New prevention control ideas	NI1	Develop a policy to restrict access except in an emergency situation (P)
	NI2	Immediately implement a regularly scheduled roof conditions visual observation plan (P)
	NI3	Design and install a monitoring system for roof crack and roof sag detection (WD)
	NI4	Repair/replace existing multipoint extensometers (WD)
	NI5	Develop a trigger action response plan (TARP) for roof movement that will initiate additional rock reinforcement installation (P)
	NI6	Design a method of stabilizing the roof within Segment 3 of the Primary Escapeway based on the information gathered from the roof monitoring program (PB)
	NI7	Consider stabilizing the adjacent Alternate Escapeway (Segment 4) to act as a buffer for securing this area (PB)
	NI8	Place charging station outside mine (EH)
New recovery measure ideas	NI9	Consider using the ventilation shaft as an Alternate Escape route (P)
	NI10	Consider installing refuge chambers in the active work areas (PB)
	NI11	Install backup generator for communication system (MH)
	NI12	Install fire detection/suppression systems on large diesel equipment (MH)
	NI13	All personnel and visitors to wear SCSRs (training needed) (PST)
	NI14	Close down charging station when the general public is underground (P)
	NI15	Finish installing lifeline in all escapeways (PST)

NI – New Ideas
EH – Eliminate Hazard
MH – Minimize Hazard
PB – Physical Barrier
WD – Warning Devices
P – Procedures
PST – Personnel Skills and Training

Several new recovery measure ideas were identified by the team. To mitigate the impact of a roof collapse, the existing ventilation shaft could be used as an Alternate Escapeway (NI9). Also, a rescue chamber could be installed in active work areas (NI10). New ideas to help recover from the fire hazard included using a backup generator for the communication system (NI11). It was also suggested that additional fire detection/suppression systems be installed on large diesel equipment (NI12). Elevated Personal Protective Equipment requirements were discussed with the goal of all personnel and visitors carrying SCSRs underground and receiving training on their use (NI13). Finally, the team suggested limiting the use of the charging station (NI14) and installing lifelines throughout all the escapeways (NI15).

5.6.3.5 – Step 5, Discuss Implementation, Monitoring and Auditing Issues

The mine's existing prevention controls and recovery measures were identified and should be monitored and audited. New potential control and recovery measures were produced in the form of an Action Plan for consideration by management. The Action Plan, with the ideas inserted, contained spaces for resources and timing of each idea. The plan was delivered to management for prioritization and implementation. Some of the new ideas are somewhat vague in that they broadly called for new designs. No attempt was made by the team to prioritize the new ideas, with the team deciding that it was inappropriate to select a specific design as part of the MHRA process. Detailed designs are not easily accomplished in an MHRA exercise. Management will need to weigh the various ideas and determine what activities are best suited for their particular circumstances.

The team seemed to agree that it was not possible to eliminate the roof fall hazard at this site. The only recourse was to decide to mitigate or tolerate the hazard. Activities to control the escapeway fire hazard were handled with well-defined actions, while activities to control roof falls relied on designs or actions that were not easily defined. Most of the new control ideas were classified as procedures (P), although new ideas were proposed in all control categories (*Figure 24*). The one control that eliminated the hazard (EH) was associated with the new idea to move the battery charging station out of the underground environment.

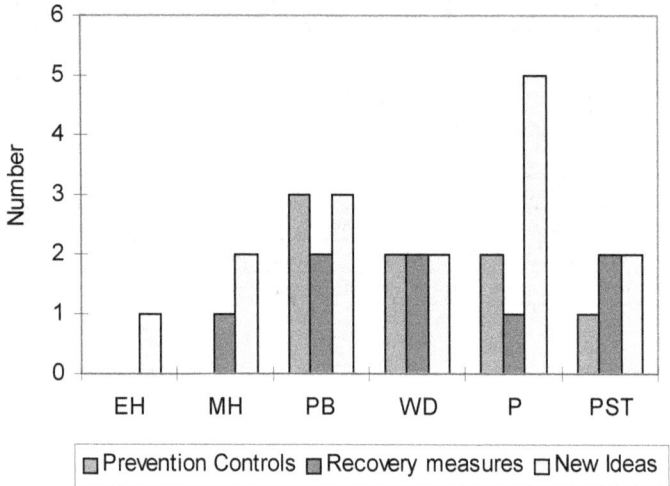

Figure 24 - Distribution of prevention controls and recovery measures for the escapeway egress blockage fire risk assessment.

5.7 – Natural Gas Ingress Risk Assessment Case Study

This case study involves two mines (Mines Ga and Gb) operating in an evaporate deposit at depths ranging from 400 to 1,500 ft using continuous miners to develop a room-and-pillar mining layout. *Figure 25* shows one level within Mine Ga, depicting main entries and production panels. The panels are mined with a higher extraction ratio. After initial crushing, the mined ore is moved to shafts by conveyors and hoisted to the surface. Several thousand feet below the mining levels lies a widespread nature gas and oil producing formation. This formation has been extensively drilled within and in proximity to the study mines for many decades to recover the natural gas and oil. The mine is aware of the current and past drilling. Significant precautions have been undertaken to reduce the likelihood that gas or oil from the reservoir below the mine does not enter the mine workings. However, because the consequences could be so significant, this mining operation has decided to review issues related to natural gas ingress using a systematic risk assessment method.

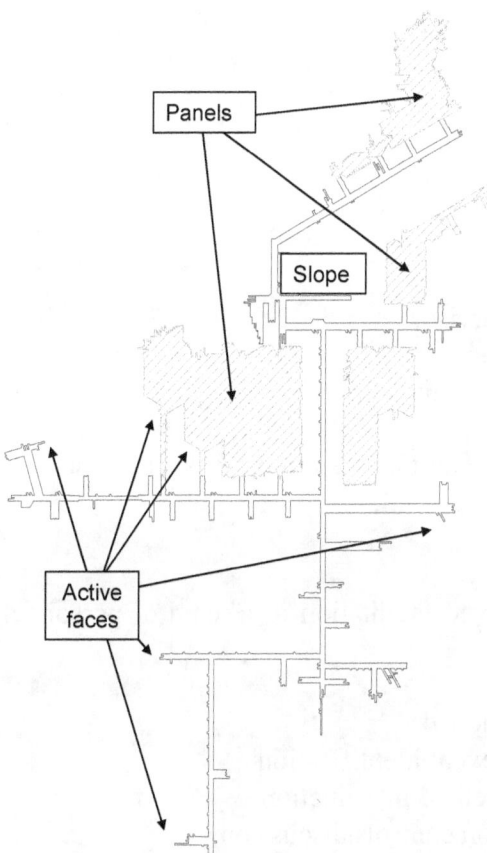

Figure 25 - Mine Ga layout showing the location of active faces, the slope and panels within one active level.

5.7.1 - Risk Assessment Scope

The objective of this risk assessment was to:
1. Review hazards associated with the potential for a natural gas inundation of the active mining area,
2. Evaluate strategies and techniques for management of the hazards, and
3. Provide information to help develop an inundation risk management plan for this mine.

The risk assessment project was scoped during discussion with the study mine's management and NIOSH personnel. The mine's personnel did not attend the NIOSH-sponsored Minerals Industry Risk Management Seminar.

5.7.2 - Risk Assessment Team

The risk assessment team was made up of persons familiar with the mine's operation and work at the mine in various capacities, as well as an external facilitator and two NIOSH observers, as follows:
- Manager of mines
- Chief mine engineer
- Manager of engineering
- Manager of safety
- Manager of operations
- Relief electrical supervisor
- Two NIOSH Observers
- Facilitator – MISHC (University of Queensland)

No representatives from labor participated in the risk assessment.

5.7.3 – Risk Assessment

The Risk Assessment involved facilitation of a team of personnel through a structured process involving the following steps:
1. Hazard description
2. Pathways identification
3. Potential unwanted event identification
4. Bow Tie Analysis method introduction
5. Causes and prevention controls discussion
6. Consequences and loss reduction controls discussion
7. Repeat of Steps 5 and 6 for all the unwanted events identified in Step 3.

Two days were dedicated to the risk assessment.

5.7.3.1 – Step 1, Identify and Characterize Major Potential Mining Hazards

The first step in the risk assessment involved identifying and understanding the hazard related to natural gas and oil around the current and planned mining operations. The primary inundation hazard was seen to be the natural gas reservoir located approximately 10,000 feet below the surface, as well as the gas in any natural or man-made conduits from below. There was also a hazard related to gas being piped on the surface or in proximity to the mines.

Pathways from the gas reservoir and gas wells into the mine were identified by the team (*Figure 26*) as follows:
- Up the inside of the drill pipe
- Up the drill hole but outside the casing
- Up/along faults
- Up/along igneous intrusions such as dikes
- Up through permeable ground along fractures or cleats
- Up collapsed breccia pipes
- Down to mine workings from wells damaged by subsidence

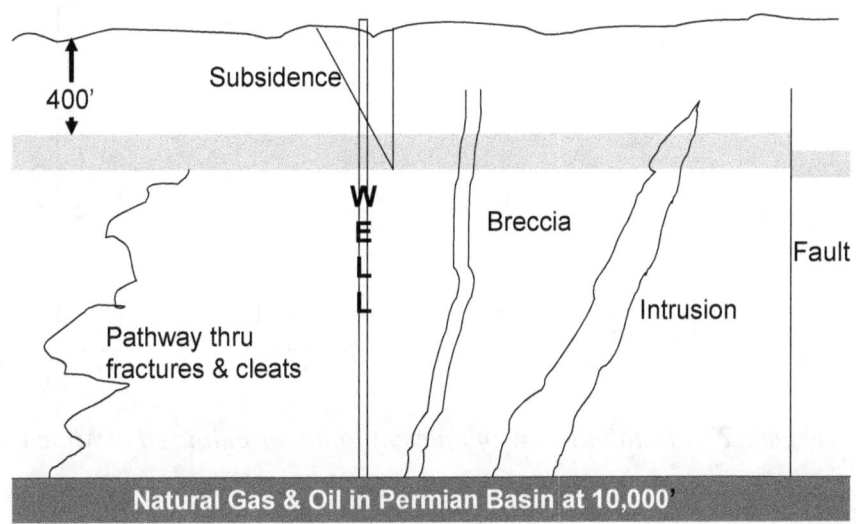

Figure 26 - The risk assessment team created this schematic to illustrate the hazards and pathways related to the study mines.

The team noted that the strata at the study mines can produce sparks during production with the cutting machine. Also, there is no intrinsically safe mining equipment at either mine. The mines are currently classified by MSHA as Class IV. Therefore, should natural gas enter the mine, heat sources could be readily available to cause an ignition.

The mine has protocols for different gas methane levels detected by hand-held units:
 At 0.5% ventilate and retest; requires notification of supervisor.
 At 1.0% ventilate, shut down equipment, notify supervisor, power off in the panel.
 At 2.0% ventilate, shut down equipment, remove personnel, and notify supervisor.

The monitors are intended for methane but will respond to natural gas, though it was not clear whether the monitors have been calibrated to the natural gas. If an explosion occurred, the related over-pressure conditions would probably destroy the brattice (cloth) stoppings. This could disrupt the mine's ventilation and hinder egress from the mines. Both mines have series ventilation where the exhaust from one panel becomes the intake air to the next panel. *Figure 27* shows the nature of Mine Ga's ventilation system where air travels in one long circuit through all the active advancing faces and production panels. If a gas ingress event occurs, the fresh air intake downstream from the gas entry point could be compromised and adversely affect egress from the mine. In addition, barricading may not work for this type of event, since miners will need to get out of the mine quickly due to the explosive potential of released gas.

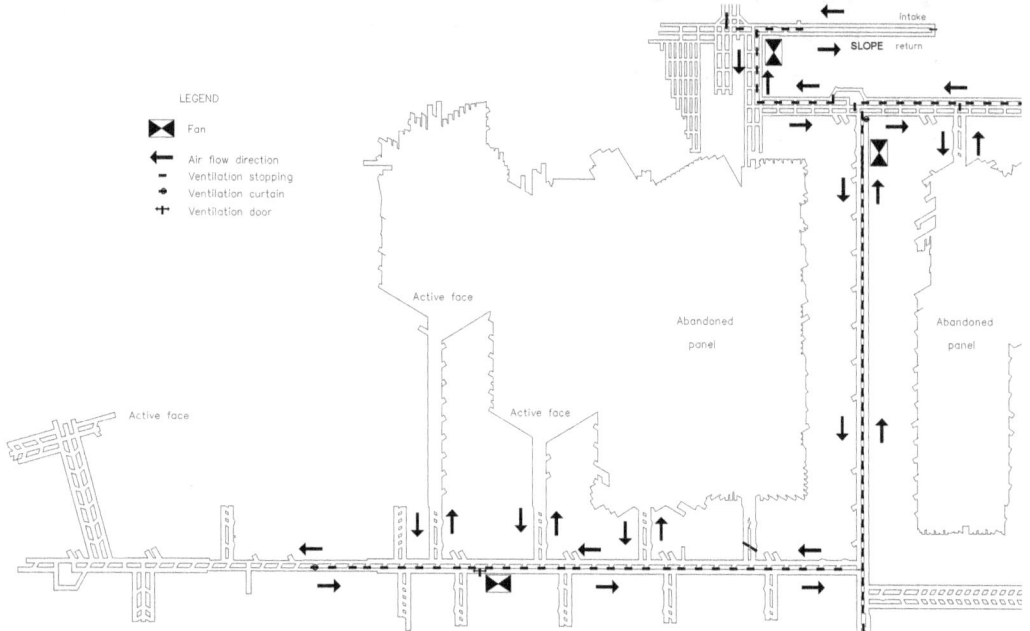

Figure 27 - Detailed view of the ventilation circuit used at Mine Ga.

The team identified that ingress of natural gas into the mine workings could have all or some of the following consequences:
- Oxygen displacement
- 2–15% hydrocarbon (methane mostly) explosive range atmospheres
- Large prolonged ignition
- Multiple fatalities
- Major mine damage
- Classification of mine to an MSHA gassy mine, which could potentially result in the closure of the mine due to higher operating cost for permissible equipment.

5.7.3.2 – Step 2, Rank the Potential Unwanted Events

The risk assessment team identified that any ingress event could be catastrophic to the mine and, potentially, mine personnel. It was therefore necessary to analyze all potential unwanted events and not to attempt to rank these events since all of them could potentially produce catastrophic consequences. In this case study, it was not necessary to perform a WRAC or a PHA to rank the unwanted events.

The team identified a list of potential unwanted events for further consideration where natural gas ingress was caused by:
1. Mining into an existing oil/gas well
2. Mining into old workings that contain gas from pathways or another source
3. Mining into fault / dike / breccia pipe which contains gas
4. Surface drillers accidentally fracturing a well in a way that creates a pathway for gas into mine workings
5. A sudden collapse in an old mined area leading to a puff or blast of gas into mine workings
6. Gas leaks into mine workings from an oil/gas well through strata
7. Gas leaks into mine workings from faults or dikes
8. Gas leaks into mine workings from breccia pipes
9. Gas leaks into mine workings through permeable ground
10. Gas leaks into mine workings from subsidence around well area
11. A gas line on the surface rupturing and the gas being sucked into mine surface ventilation intake (note that the fans are underground).

<u>5.7.3.3 – Step 3, Determine Important Existing Prevention Controls and Recovery Measures</u>

The BTA approach was used to consider each of the above events. For example, the No. 1 potential unwanted event "Mining into an existing oil/gas well" forms the top event in the center of the bow tie of *Figure 28*. For each potential cause on the left side of the bow tie, the risk assessment team identifies both existing and new key prevention controls. Next, the right side of the bow tie is acted on and existing and new potential recovery measures are identified. This process is repeated for each of the potential unwanted events.

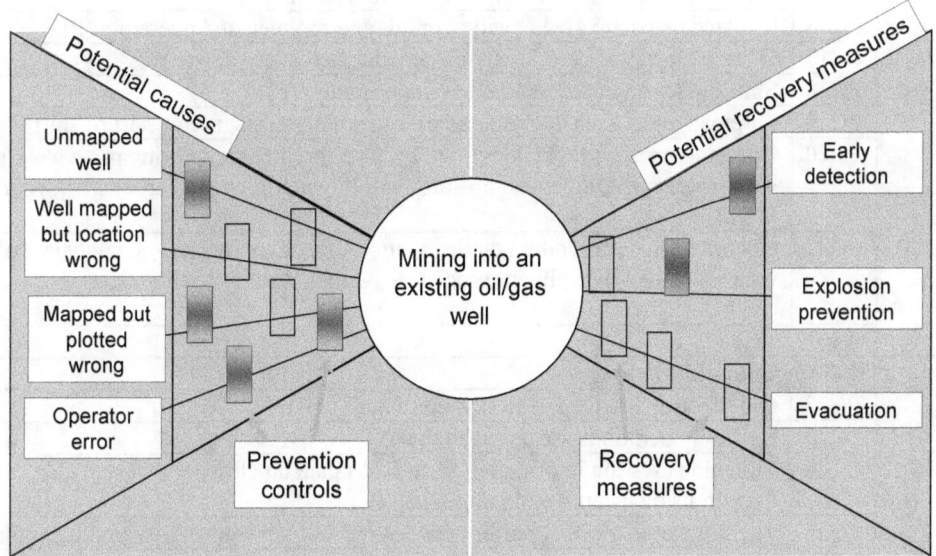

Figure 28 - BTA for mining into an existing oil/gas well top event. Potential causes are listed on the left side of the bow tie with potential recovery measures on the right.

Based on the analysis that was done by the team, 33 key existing controls were identified as being in place throughout relevant phases of the mining operations (*Table 33*). As such, these controls should be reinforced, monitored and audited with priority. Controls focused on reducing potential well misidentification, reducing operator errors, developing more accurate geologic information of discontinuities that could transmit oil and gas, and ventilating abandoned areas of the mine that might contain oil and gas. BTAs were needed for only the first four potential unwanted events. Previously identified controls adequately covered the remaining potential unwanted events.

Table 33 – Existing key prevention controls for the natural gas ingress risk assessment (left side of the BTA).

Top Event => (1) Mining into an existing oil/gas well		
Cause		**Existing controls to prevent the unwanted event**
Unmapped well	PC1	Physically checks surface well location, i.e. topographic, aerial, recon, etc. (P)
	PC2	Check well location against well location supplied by the State (P)
	PC3	Monitor for new application for drilling (APD) (P)
	PC4	Check on the ground for old well location by locating old bricks, oil seeps, etc (P)
	PC5	Develop and update oil and gas map database (P)
Well mapped but location wrong	PC6	Draw a 1,320-ft circle around oil wells and a 2,640-ft circle around gas wells (PB)
	PC7	Resurvey all locations on mine property (GPS survey) and compare to mine survey to reduce likelihood of wrong location on map. (P)
	PC8	Quality Software (CAD and Geographic) check to compare each answer (P)
	PC9	Third party check of data and map location (P)
Mapped but plotted wrong	PC10	Check ongoing surveys (P)
	PC11	Keep elevation of all mine workings in mine panels (P)
	PC12	Use of best technology and survey every shot with triple flop of scope (P)
Operator error	PC13	Plot oil and gas locations in critical mining areas on escape maps located in dinner hole (P)
	PC14	Locate oil and gas wells with plot lease boundaries on foreman regular work area maps (P)
	PC15	Panels are surveyed every 125 feet to fix locations (P)
	PC16	Updated maps are given every work day to foremen and operators (P)
	PC17	Work expectations are discussed and clarified every day (P)
	PC18	Shift bosses have access to maps underground (P)
	PC19	Foreman gives each operator his personal map on surface at start of shift (P)
	PC20	Operator error would be identified by shift foremen monitoring mine work to ensure that well location is known and operator is following mine plan and matches survey location (PST)
	PC21	Operator error (depending on the nature of the inappropriate behavior) would be dealt with by known disciplinary procedure (verbal, written communication, days off, fired) (PST)
	As above: see PC6	
Top Event => (2) Mining into old workings		
	PC22	Old mines in area are naturally ventilated (MH)
	PC23	Existing operations are unsealed (MH)
	PC24	The mine uses the information from BLM data on closed mines and also maintains good data on its own captive old mines (PB)
	PC25	The mine uses a 100-ft buffer between old and new mines included in all plans (PB)
	PC26	There is positive air pressure from ventilation for active mining areas causing gas to migrate toward old workings, unless old workings have a higher pressure (MH)

Top Event => (3) Mining into a fault / dyke or breccia pipe		
	PC27	Breccia pipes and dikes have been mapped and located but all faults have not. The area is geologically stable so there are not too many faults. The ore horizon is some twelve different horizons over a 400-foot depth. Some horizons are economic but some are not (P)
	PC28	Gas inrush hazard would be considered to be a slow leak type of event and not a major inrush of gas (MH)
	PC29	The mine has past experience with mining into breccias area with an active oil seep (PST)
Top Event => (4) Drillers accidentally hydra-fracturing a well in a way that creates a pathway from gas into mine workings		
	PC30	Some drillers notify State Oil Conservation District (OCD) and some companies call the mine (PST)
	PC31	Some events are mandated to be reported by the OCD (State) but that requirement only affects ~ ten % of the land. Most land is owned by BLM. (P)
	PC32	A permit to drill is required which allows the mine to publicly comment on the drilling application (P)

PC – Prevention Controls
EH – Eliminate Hazard
MH – Minimize Hazard
PB – Physical Barrier
WD – Warning Devices
P – Procedures
PST – Personnel Skills and Training

Based on the analysis that was done by the team, 15 key recovery measures controls were identified (*Table 34*). These recovery measures fit into three categories: early detection, explosive prevention, and evacuation. Early detection relies on supervisors monitoring the working environment. Explosion prevention requires numerous actions to de-energize the mine or alter ventilation. Evacuation is focused on communicating instructions to the workforce and assisting in the movement of workers out of the mine with breathing equipment. A BTA was only needed for the first potential unwanted events. All other unwanted events produced the same set of recovery measures.

Table 34 – Existing key recovery measures for the natural gas ingress risk assessment (right side of the bow tie).

Top Event => gas inrush occurs		
Early detection	RM1	The mine has ongoing gas detection for O_2 and CH_4 for operators and O_2, CO and CH_4 for all supervisors (WD)
Explosion prevention	RM2	If major inrush of gas occurs the expectation is that all equipment power will be shut off (miner can trip power to transformer from mining machine), supervisors are notified and all other equipment would be shut down (P)
	RM3	Power can be shut off to rest of mine except hoist by the electrical power supplier (P)
	RM4	Mine power can be shut down by surface personnel (P)
	RM5	Ventilation fans are underground and can be shut off or reversed (MH)
Evacuation	RM6	Current evacuation plans call for mine workers to utilize diesel equipment to get to shaft through the intake auxiliary escape route (PST)
	RM7	Escape and emergency egress is practiced every six months (PST)
	RM8	Other U/G personnel not in the immediate vicinity of gas inrush would be notified by one of three ways: page phone system, word of mouth, or flashing lights on belts (WD)
	RM9	The hoistman is the key communication person (P)
	RM10	Workers when they call in are directed to which egress pathway to take (P)

	RM11	The person calls his supervisor who may or may not know which way to escape. Supervisor may or may not be in communication with hoistman (P)
	RM12	At both mines power of hoist is totally isolated from mine power (MH)
	RM13	There are refuge chambers at each mine fed by compressed air from surface with enough food, water and air for 80 people (PB)
	RM14	There are caches of SCSR breathing apparatuses located at each mine (one hour units) (PB)
	RM15	Miners carry the ten-minute Ocenco oxygen in addition the W65 CO units (PB)

RM – Recovery Measures
EH – Eliminate Hazard
MH – Minimize Hazard
PB – Physical Barrier
WD – Warning Devices
P – Procedures
PST – Personnel Skills and Training

5.7.3.4 – Step 4, Identify New Prevention Controls and Recovery Measures

New ideas were identified by the team during the risk assessment to further reduce the ingress risks at the mine (*Table 35*). Three new prevention controls were focused on identifying well location (NI1 to NI3). Three ideas dealt with the potential for natural gas to be retained in abandoned panels (NI4 to NI6). Management will need to weigh the advantages of sealing old works as recommended in NI6 against the current conflicting prevention controls (PC22 and 23) where the old works are not sealed. Three other ideas attempt to influence the practices of future drilling operations near the mines (NI7 to NI9). Seven new recovery measure ideas were developed by the risk assessment team. Three ideas focused on improving early detection of natural gas ingress (NI10 to NI12). Two ideas dealt with explosion prevention (NI13 and NI14) and two with evacuation issues (NI15 and NI16).

Table 35 - New ideas for mitigating risk of natural gas ingress.

Prevention Controls	
NI1	Check the mine survey every 2000–3000 feet of advance (P)
NI2	Reinforce the need to turn on the AutoCAD layers to show well locations (PST)
NI3	Identify well locations before any new panel is planned or mined (P)
NI4	Consider drilling from either surface or underground to classify, locate and determine gas pressure of old mines suspected to be in the area of active mining. There is a possibility that some existing extracted panels ventilated by the mines might have some accumulation of gas. The critical panels are the panels which are 20 to 30 years old. (MH)
NI5	Consider drilling and investigating existing extracted panels which are greater than some value in time (to be established by mine personnel based upon experience) (MH)
NI6	Seal old mine workings to keep any accumulated gas in the old workings area even if a fall of ground occurred in the old workings (PB)
NI7	Increase efforts to eliminate drilling from surface mine property in close proximity of mine and get drillers to utilize directional drilling (EH)
NI8	Influence the drillers or third parties to get drillers to utilize better drilling methodologies (MH)
NI9	Increase mine awareness of any drilling problems by either a cooperative effort with drilling companies, OCD (State), BLM (Federal), or combination of all the above. This needs to include gathering and communicating vital information on drill location, well type, and all other relevant data (P)

	Recovery Measures
NI10	Use new communication system to add selected gas monitoring locations at various places in the mine (WD)
NI11	Add new monitoring sensor to new communication system at all intake shafts or all shafts to shut down fans if gas leak detected. (Mine personnel will determine gas trips applying experience, consequences, and all possibilities to determine trigger values on all gas monitors at the location that must be monitored. Trigger values may be different depending on location and impact to and possible consequence) (WD)
NI12	Investigate the use of sampling pump technology to test atmosphere closer to face than the current continuous miner operator by placing sensor technology on CM cutter head. Same preset values are present for triggering action items for the gas values at 0.5%, 1% and 2% gas (WD)
NI13	Examine the use of blast doors or isolation doors underground to reduce ignition consequences and isolate any event to that section of the mine (PB)
NI14	Consider methods for automatically dropping power in the section where gas inrush occurs (MH)
NI15	Consider the role of the initial event communicator and that person's capability and critical decision-making, and include in the Emergency Response Plan (ERP). At one of the mines, there is an additional need because there is no cager, just a hoistman (P)
NI16	Consider all the factors and develop understanding of all tradeoffs of egress options for series ventilation and the inexperience of the new mine workers. At the present time, even though both mines are connected, the company has not practiced egress from one mine to another (PST)

NI – New Ideas
EH – Eliminate Hazard
MH – Minimize Hazard
PB – Physical Barrier
WD – Warning Devices
P – Procedures
PST – Personnel Skills and Training

These ideas for new potential prevention controls and recovery measures should be addressed through the development of an Action Plan. Assuming that the information provided in the risk assessment was accurate, completion of the Action Plan and an increased focus on monitoring and auditing of the key identified controls would appear to provide an opportunity to effectively reduce the risk of fatalities related to gas ingress at the case study mines.

<u>5.7.3.5 – Step 5, Discuss Implementation, Monitoring and Auditing Issues</u>

The risk assessment team had a wide range of expertise familiar with the natural gas ingress hazards and the associated risks to the mining operation and underground workforce. The team was well-represented by key management personnel who had the authority, responsibility and experience necessary to support an MHRA. The team acted as a cohesive unit who cared deeply about all the employees at the mine but also as visionary people who could think outside the normal everyday existence at the mine. All scenarios were evaluated and completely discussed by all mine personnel until everyone involved in the exercise felt the matter had been completely evaluated. However, the team lacked representation from labor and outside expertise.

The list of key existing prevention controls and recovery measures demonstrates that the mining operation has spent considerable energies thinking about this major hazard. But it is equally apparent from the large number of solid new ideas that more could be done. Some of the ideas were very practical and had a high potential for being implemented. Others seemed more difficult and involved the actions of outside government agencies. While it is less likely that these ideas could be implemented by the local mining operations, it is possible that others agents in the company could help. This could be an example where units lower in an organization's structure influence the actions of units higher in the organization's structure through the MHRA process.

At Mine G, there is a strong reliance on prevention controls (PC) classified as procedures (P) (*Figure 29*). The high reliance on procedures increases the potential for human error to play an important role. The new ideas were more evenly spread over the different control categories.

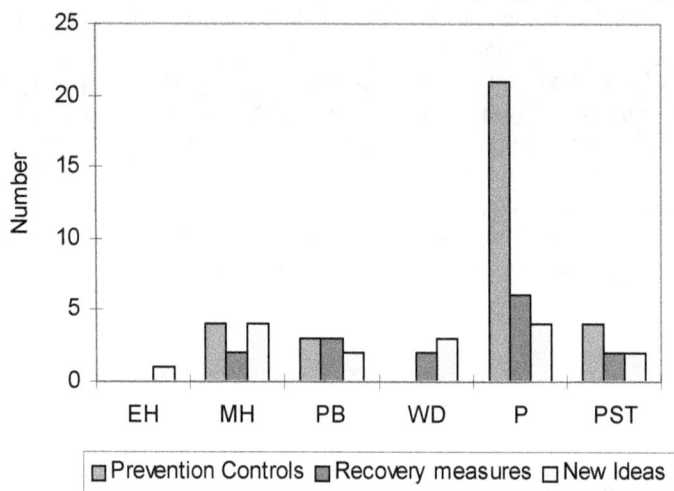

Figure 29 - Distribution of prevention controls and recovery measures for the natural gas inundation risk assessment.

5.8 – Conveyor Belt Fire Risk Assessment Case Study

Mine H is an underground room and pillar coal mine with rooms 48 to 54 inches high by 18 feet wide. The mine employs approximately 100 miners and operates three mining units with typical equipment such as continuous miners, shuttle cars and a conveyor belt system extending from three different working faces to the surface (*Figure 30*). The operator did attend a NIOSH-sponsored training class but had expressed a desire to participate in the pilot project.

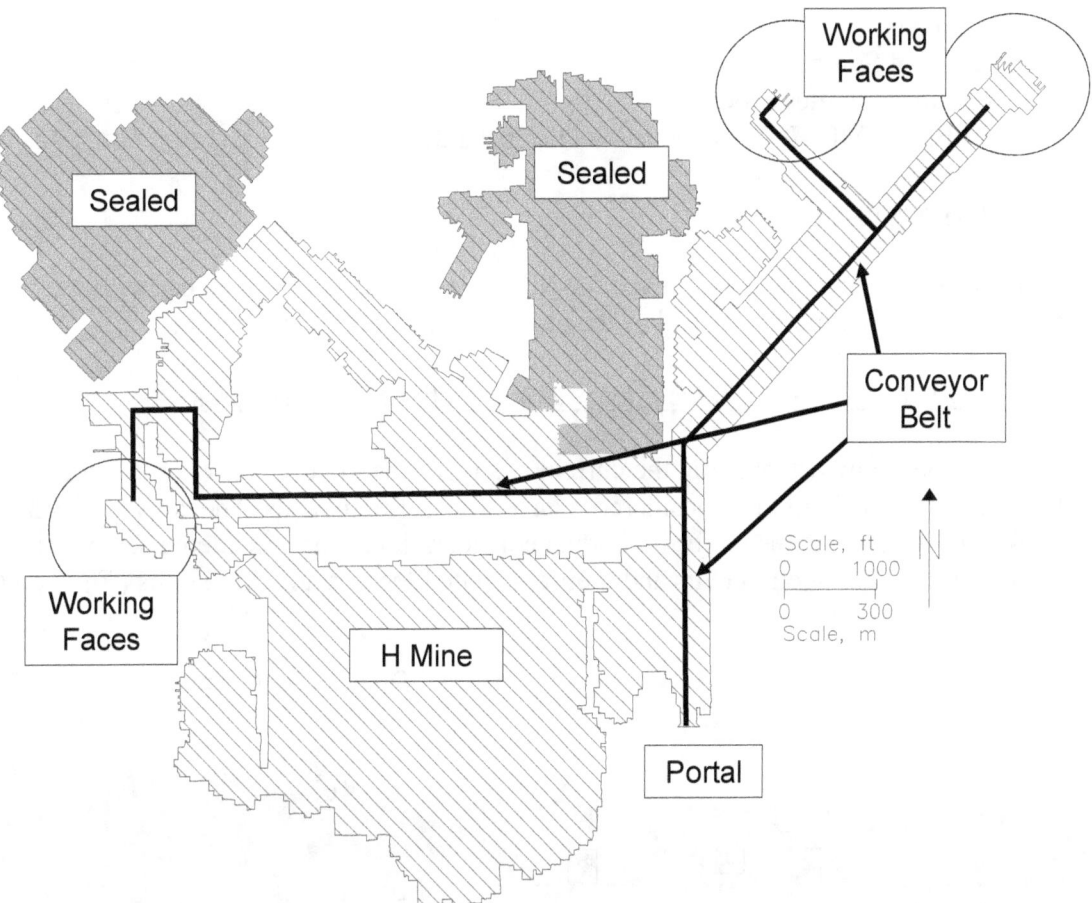

Figure 30 - Mine H layout showing the location of the conveyor belt and working faces.

5.8.1 - Risk Assessment Scope

The objective of this risk assessment is to 1) review major hazards associated with fire potential on underground conveyor belts at the mine, 2) evaluate fire prevention strategies, early detection techniques, primary fire suppression systems, fire fighting techniques, and mine evacuation procedures in the context of these hazards, and 3) develop a major hazard management plan for this mine site. The mine uses belt air to partially ventilate the working faces. Some controls required by MSHA regulations are: 1) intake air monitor at the outby end of each section, 2) Carbon Monoxide (CO) monitors at specific intervals along the belt, and 3) a fixed fire protection, water deluge system at each drive set to trigger at 165° F. The mine was interested in examining additional controls to lower the risk of fire on its underground conveyor belt system.

5.8.2 - The Risk Assessment Team

The risk assessment team was made up of persons employed at Mine H as well as from its parent company. The team members included:
- Mine superintendent
- Shift foreman
- Two miners - underground and outside supply
- Engineer
- Electrician
- Director of Safety
- Two subject matter experts
- Facilitator – MISHC (University of Queensland)

5.8.3 - Risk Assessment

This risk assessment case study followed the MHRA approach as outlined earlier but used a three-dimensional risk matrix instead of the more common 5 x 5 risk matrix.

5.8.3.1 - Step 1, Identify and Characterize Major Potential Mining Hazards

This exercise began by establishing the current design of the mine's conveyor system and identifying risks that should be considered related to belt fire hazards. The conveyor belt system was broken down into segments for individual consideration. The considered segments consisted of individual section belts and their associated feeders and drive units (*Figure 31*).

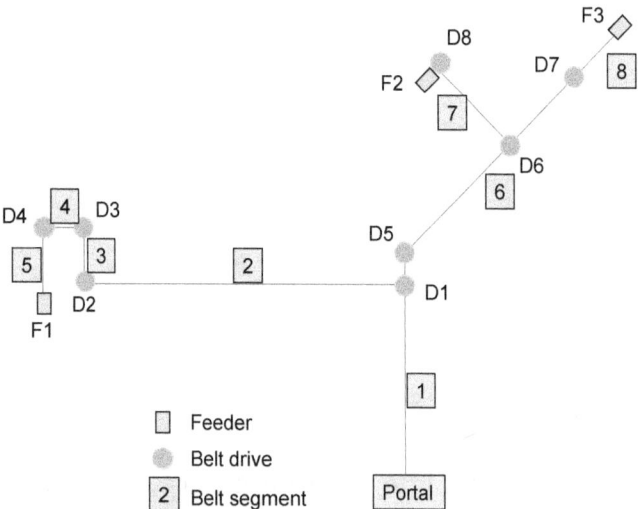

Figure 31 - Segments of the conveyor belt system.

Eight individual conveyor belt segments were identified with the following characteristics (*Table 36*).

Table 36 - Characteristics of eight conveyor belt segments.

Segments	Characteristics
1	3,100 ft of 42-inch-wide belt from portal to drive D1 with an air velocity of 490 ft/s,
2	5,375 ft of 36-inch-wide belt from drive D1 to drive D2 and containing cribbed area with an air velocity of 170 ft/s,
3	900 ft of 36-inch-wide belt from drive D2 to drive D3,
4	700 ft of 36-inch-wide belt from drive D3 to drive D4,
5	1,200 ft of 36-inch-wide belt from drive D4 to feeder F1 with an air velocity of 120 ft/s,
6	4,580 ft of 36-inch-wide belt from drive D1 to drive D7 and containing drives D5 and D6 with an air velocity of 220 ft/s,
7	1,820 ft of 36-inch-wide belt from drive D6 to feeder F2 and containing drive D8 with an air velocity of 90 ft/s,
8	910 ft of 36-inch-wide belt from drive D7 to feeder F3 with an air velocity of 220 ft/s.

The team then listed the types of related hazards that should be considered in the assessment. Fuel and heat sources are identified in *Table 37*.

Table 37 - Fuel and heat sources along the conveyor belt.

Fuels	Heat sources
Coal dust	Electricity
Timber	Friction (rollers, belt, bearings, etc.)
Grease	Welding
Paper	
Rubber (belt)	
Hydraulic Fluid	
Insulation on wires	
Plastic pipe	
Canvas	
Methane (at conveyor dump points)	

5.8.3.2 - Step 2, Rank Potential Unwanted Events

Once the conveyor segments and hazards had been listed, the team was ready to identify the potential unwanted events for the entire conveyor belt system. The list of potential unwanted events was ranked for risk using a WRAC. The team considered each conveyor segment and each fire threat individually in order to systematically identify potential unwanted events. Forty-one different potential unwanted events were identified *Table 38*.

Table 38 – Potential unwanted events for the entire conveyor belt system

#	Location	Unwanted Event
1	Feeder	Bearing failure causes fire in coal dust
2	Feeder	Electrical short causes fire in coal dust
3	Feeder	Friction, hot motors that are above normal operating temperature set fire to wood cribs
4	Feeder	Drive chain heat sets fire to oil/grease/coal build-up and sets fire to the drive chain
5	Feeder	Rocks/metal bits in feeder cause sparks leading to a fire
6	Feeder	Rescuer contents exposed to air when broken in feeder, starting an exothermic fire
7	Section Belt	Tail roller failure (back from face) generates heat and ignites coal dust/grease
8	Section Belt	Misaligned belt generates friction (when it stops) and starts fire
9	Section Belt	Belt clearance problems cause rubbing and fire
10	Section Belt	Structure (e.g. tail piece) rub on belt causing fire
11	Section Belt	Structure/metal (e.g. rollers and structure) friction causes fire
12	Section Belt	Belt causes electricity fault on belt, (Jabcos) 110V causes fire
13	Section Belt	Electrical fault on HV crossovers for various purposes causing fire
14	Section Belt	Belt box failure /fault causes fire
15	Section Belt	Welding on belt structure leads to fire
16	Section Belt	Improper installation of main belt structure (hangers) leads to friction and fire
17	Section Belt	Structural failure (hung belt failure) leads to spillage/damage/friction and fire
18	Drive	Electrical fault in drive causing heat and fire
19	Drive	Scrapers wear to a point where they are metal-to-metal causing friction and fire
20	Drive	Misalignment of scraper causes build-up that leads to friction, heat and fire
21	Drive	Misalignment of belt causes "strings" that get heated up and catch fire
22	Drive	Welding at drive causes fire
23	Drive	Floor heave misaligns drive causing friction, heat and fire
24	Drive	Malfunction on the slip (belt control error) unit causes heat and fire
25	Drive	Batteries fault while driving a mantrip and catch fire
26	Drive	Bearing faults at drive generate heat and fire
27	Drive	HV box fault on belt cause fire
28	Drive	Rock from roof falls into drive area causing friction and fire
29	Drive	Hydraulic brake slip generates heat and if fluid leak then fire
30	Special	Fire starts in cribbed area of Segment #2 due to typical belt fire reasons (more spillage, harder to inspect, harder to clean)
31	Special	Float dust/trash around air locks on belt leads to fire or NOT hot spots
32	Special	Belt fire in Segment #1 has major impact on inby ventilation
33	Special	Belt fire in Segment #6 affects ventilation intake undercasts, compromising supply inby
34	Special	Section #7 transfer point roof conditions leads to rock/friction/fire
35	Special	Fire at undercast, Section #1 changes ventilation
36	Special	Fire at overcast, Section #1 changes ventilation
Unusual Fuel sources		
37	Feeder	Greasy rags, paper, housekeeping problems cause fire
38	Section Belt	Unnecessary fuels (greasy rag/poor housekeeping leads, coal, timber, etc.) to fire
39	Drive	Housekeeping problems lead to fire
40	Drive	Canvas fire starts
41	Drive	Oil sump for chain catches fire on drive

The team applied a three-dimensional, subjective risk matrix (*Table 39*) to identify priority unwanted conveyor fire events for further consideration. This method involved selecting, for each identified unwanted event, the possible Maximum Reasonable Consequence (MRC) of that event, the Most Likely Consequences (MLC) of that event, and the likelihood of that event occurring. The subjective ranks for each event were then defined. *Table 40* shows the 12 highest risk unwanted events.

Table 39 – Three-dimensional risk ranking method used at Mine H.

CONSEQUENCE MATRIX	Most Likely Consequence (MLC)				
Maximum Reasonable Consequence (MCR)	MFF**	Almost MFF	Serious Fire	Minor Fire	No Fire
MFF	A	A	B	C	D
Single fatality fire	A	A	B	C	D
Serious LTI*	A	B	C	D	E
Avg LTI	B	C	D	E	E
Minor LTI	C	D	E	E	E
RISK RANK MATRIX	Likelihood of Occurrence				
From the above Consequence Matrix	5 - Common (>1 per week)	4 - Likely (1 per month)	3 – Moderate (1 per year)	2 - Unlikely (1 per several years)	1 - Very Unlikely (almost never)
A	1	2	4	7	11
B	3	5	8	12	16
C	6	9	13	17	20
D	10	14	18	21	23
E	15	19	22	24	25

LTI* = lost-time injury
MFF** = multiple fatality fire

Table 40 – The highest priority risks identified by the WRAC.

#	Unwanted Event	MRC	MLC	C	L	R
1	Fire starts in cribbed area of Segment #2 due to typical belt fire reasons (more spillage, harder to inspect, harder to clean)	5	4	A	4	2
2	Float dust/trash around air locks on belt leads to fire or NOT hot spots	5	3	B	4	5
3	Belt fire in Segment #1 has major impact on inby ventilation	5	3	B	4	5
4	Fire at overcast, Section #1 changes ventilation	5	3	B	4	5
5	Electrical fault on HV crossovers for various purposes causing fire	5	3	B	3	8
6	Large structural belt failure leads to spillage/damage/friction and fire	5	3	B	3	8
7	Floor heave misaligns drive causing friction, heat and fire	5	3	B	3	8
8	Malfunction on the slip (belt control error) unit causes heat and fire	5	3	B	3	8
9	Belt fire in Segment #6 affects ventilation intake undercasts, compromising supply inby	5	3	B	3	8
10	Fire at undercast, Section #1 changes ventilation	5	3	B	3	8
11	Structure (e.g. tail piece) rub on belt causing fire	5	2	C	4	9
12	Structure/metal (e.g. rollers and structure) friction causes fire	5	2	C	4	9

The highest ranked risk from the WRAC was a fire starting in the cribbed area of conveyor belt Segment No. 2 (*Figure 32*). Because a fire at this location represents the highest risk to the mine, more time was dedicated to discussing controls for this unwanted event.

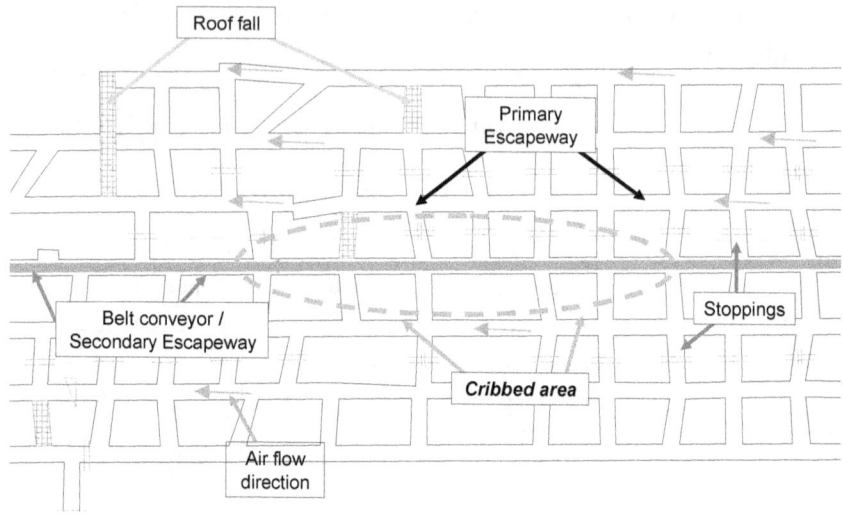

Figure 32 - Conditions within the cribbed area of conveyor belt Segment #2.

The cribs were placed in this area to help prevent the convergence of the roof and floor. Several roof falls had occurred in adjacent entries and a relatively large area was being subjected to excessive pressure that was attempting to force the entry closed. Under these conditions, a standard control practice is to support the entry with standing structures to resist the roof-to-floor closure. Wood cribs are often used for this purpose. Drawbacks for this control include: 1) reduced access to the area, 2) increased air velocity as the cross-sectional area of the entry is effectively reduced, and 3) elevated sources of fuel (wood) to the area.

<u>5.8.3.3 - Step 3, Determine Important Existing Prevention Controls and Recovery Measures</u>

The highest risk unwanted events (top events) identified by the WRAC were selected for a much more detailed analysis using the BTA. The BTA analyzed the control measures intended to prevent the unwanted event and all consequences leading to the unwanted initiating event. The results of the Mine H BTA are provided in Appendix B.

The key controls identified by the process as currently in place throughout the underground belt conveyor system are listed in *Table 41*. Fourteen existing prevention controls were identified and grouped into two categories. The conveyor construction controls are required by mining regulations or are considered Best Practices. The conveyor maintenance controls focus on preventive maintenance issues and good housekeeping.

Table 41 - Summary of existing prevention controls and recovery measures from a potential conveyor belt fire.

Existing prevention controls			
Conveyor construction	PC1	The conveyor is designed for the load, speed, etc. (MH)	
	PC2	The conveyor is hung straight/correctly (MH)	
	PC3	High-voltage cable crossovers are hung high over the conveyor (P)	
	PC4	High-voltage cables run in a pipe to protect cable (PB)	
	PC5	Drives are set to put dripping onto the outby belt (MH)	
	PC6	Skirting and bins at transfer points that decrease spillage (PB)	
	PC7	"Land mines, mouse traps, rabbit holes" that shut down the belt if excessive spillage is detected at transfer points (WD)	
	PC8	Canvas at drives designed to hang clear of the machinery (P)	
Conveyor maintenance	PC9	Mechanics inspect/repair each drive every day (P)	
	PC10	Belt maintenance personnel keep all belts tracking correctly (PST)	
	PC11	If a problem is found (drive seals, rollers, etc.) repairs are done (PST)	
	PC12	Spills are cleaned up on back shifts (P)	
	PC13	A designated person is assigned to fill drives with oil (P)	
	PC14	All oil spills are cleaned up (P)	
Existing recovery measures			
Fire identification and communication	RM1	Persons in the area should detect fire at feeder and other points (PST)	
	RM2	CO monitors around feeders and other points are set at 5 ppm alert and 10 ppm alarm with audio that warns surface of possible fire (tested weekly) WD	
	RM3	Persons are trained to contact the control room person if there is a suspected or actual fire (PST)	
	RM4	If a fire is suspected, the control room person shuts off the belt and starts calling the sections to alert all miners (P)	
	RM5	Persons in sections would notice the stationary belts or hear the phone warning and/or alarm (PST)	
	RM6	Persons are trained to contact supervisor(s) and surface attendant(s) to clarify the problem (PST)	
	RM7	Persons would fight the smaller fire outby (intake side) (PST)	
	RM8	Persons fighting the fire would let surface attendant know if fire is beyond fighting (P)	
	RM9	Surface attendant would let all underground know if it is time to evacuate (P)	
	RM10	New employees are introduced to the mine and emergency procedures and persons understand procedures (PST)	
Fire fighting capacity	RM11	Fixed fire suppression and fire hoses are available at the feeder and drives (MH)	
	RM12	Every section and belt drive has a fire hose cart for quick delivery of hoses to fire location (every 15 water pipe joints there is a tap) (MH)	
Emergency escape	RM13	All persons have self-rescuers (1-hr maximum) to help them get to fresh air and ride out (PB)	
	RM14	All persons know how to get to fresh air (PST)	
	RM15	More than one intake egress is available for escape (MH)	
	RM16	Mine practices escapeway drill and fire fighting drill regularly (mock drill) (P)	

PC – Prevention Controls
RM – Recovery Measures
EH – Eliminate Hazard
MH – Minimize Hazard
PB – Physical Barrier
WD – Warning Devices
P – Procedures
PST – Personnel Skills and Training

Sixteen existing recovery measures focused on fire identification and communication, fire fighting capacity, and emergency escape controls were identified. Here again these controls represent a combination of complying with mining regulations and Best Practices.

5.8.3.4 - Step 4, Identify New Prevention Controls and Recovery Measures

As the risk assessment team identified existing prevention controls and recovery measures associated with a conveyor belt fire, new ideas were proposed to help further reduce risk. Thirteen new ideas were identified, six related to prevention control and seven related to recovery measures (*Table 42*).

Two new prevention control ideas (1 and 2) focused on the #1 ranked risk – fire along the conveyor belt in the cribbed section of Segment #2. The combination of using infra-red cameras and thermometers to detect hot spots and better control of coal fines were viewed as significant controls to further mitigate risk. New idea 3 focused on eliminating the wood fuel supply from future sites where roof-to-floor convergence might occur. The NIOSH program STOP (Support Technology Optimization Program) is a design tool that can be used to help investigate different supplemental support options (http://www.cdc.gov/niosh/mining/products/product99.htm). New idea 4 focused on future conveyor construction, proposing to eliminate many of the hazards through better construction techniques. New idea 5 relies on an SOP to ensure that cable splices are done to standards. New idea 6 deals with a relatively rare phenomenon – an SCSR mistakenly entering the conveyor belt feeder. This has happened only once at this mine, but the incident resulted in a hot fire that lasted for many minutes.

Table 42 – New ideas proposed by the risk assessment team for preventing or recovery from a conveyor belt fire at Mine H.

New prevention control ideas	NI1	Conveyor Monitoring / Inspection - combine the role of examining and cleaning the belt (P)
	NI2	Investigate infra-red cameras and thermometers to detect hot spots (WD)
	NI3	Conveyor Cleaning – install a water valve outby the cribbed area, have weekly wash downs (P)
	NI4	Install a knee wall to deflect water and collect fines (MH)
	NI5	Use hydraulic jacks rather than wooden cribs as supplemental support when roof-to-floor convergence is a problem. Consult NIOSH "STOP" program for assistance (MH)
	NI6	Conveyor Construction – survey and mark drives for conveyor hanging, plan HV crossovers, and cut bottoms in new drive areas (MH)
	NI7	Conveyor maintenance - ensure splices are square, complete an SOP for splice inspection (P)
New recovery measure ideas	NI8	Self-Rescuers – communicate the related fire source risk and provide locations on equipment for rescuers (PST)
	NI9	Fire Fighting Plan – review the fire fighting plan to ensure the key actions are understood, developing a control room check list for actions; define the MSHA interaction and consider an emergency info "sticker" for personnel (P)
	NI10	Fire Identification - supplement CO monitoring with smoke monitoring in belt headings (WD)
	NI11	Fire Communication - investigate technology to notify persons to leave the mine (WD)
	NI12	Fire Fighting Capacity - analyze fire fighting capability and hang ribbons on belt line for fire taps and hose locations (P)

	NI13	Fire Escape – train all to use optional alternate (3rd emergency egress) (PST)
	NI14	Specific to the Cribbed Area in Segment #2- install additional phone and fire suppression over the conveyor belt in this area (MH)
	NI15	Fire Event Simulation - use event simulation to test response to fire (MH)

NI – New Ideas
EH – Eliminate Hazard
MH – Minimize Hazard
PB – Physical Barrier
WD – Warning Devices
P – Procedures
PST – Personnel Skills and Training

The seven new recovery measure ideas covered a range of emergency response issues associated with detecting a fire (NI8), communicating its occurrence and location (NI9), fighting the fire (NI10), and escaping safely from the mine (NI11). The team had one new idea (NI12) specific to the cribbed area in Segment #2. New ideas 7 and 13 focused on improving the existing fire fighting plan through a control room checklist and event simulations.

<u>5.8.3.5 - Step 5, Discuss Implementation, Monitoring and Auditing Issues</u>

The existing prevention controls and recovery measures are obviously keys to reducing the risk of a conveyor belt fire and, therefore, should be reinforced, monitored and audited with priority. The new ideas were compiled into an Action Plan with the recommendation that each item be evaluated within a specific time frame and a decision made by management as to which would be implemented by the mine. Lastly, a presentation was made by the risk assessment team to mine management stressing the above points.

The existing and new prevention control and recovery measures identified with the BTA fell largely within the mitigation and tolerance range of hierarchy responses to the identified hazards. The team did not identify any controls that would have eliminated the hazard entirely. If this mining process were not used at this mine, many of the risks analyzed would have been diminished; as one example, using belt air to ventilate the working faces is responsible for many of the mine's conveyor belt fire high-consequence events. It is difficult for an MHRA to consider hazard elimination when the mine is mature and the action of hazard elimination might produce other unfavorable mining conditions. The high-voltage power cables crossing the conveyor belt line is another example of the difficulty in hazard elimination actions. Here actions focused on ways of mitigating or tolerating the risks associated with this hazard.

A number of the existing controls discussed by this risk assessment team identified mitigation techniques (MH and PB, see *Figure 33*). Also, five new ideas were classified as controls to minimize hazards (MH), where some technology would independently aid in preventing a fire or minimizing the resultant losses if a fire were to occur. If a fire occurred and recovery measures (RM) were needed, there was a high reliance on procedures (P) and personnel skills and training (PST) (*Figure 33*).

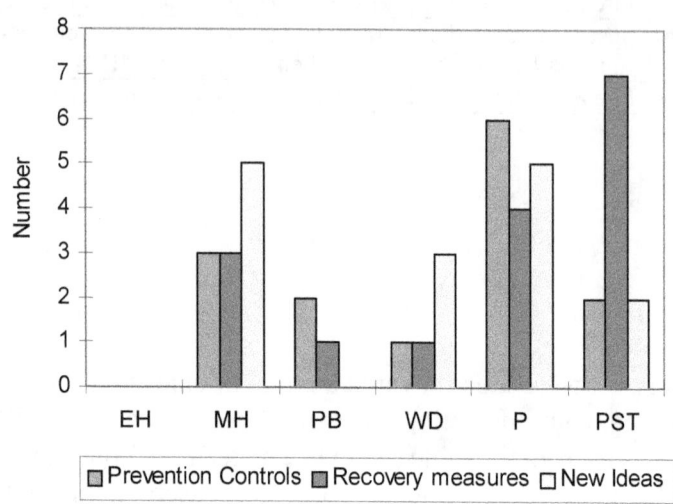

Figure 33 - Distribution of prevention controls and recovery measures for the conveyor belt fire risk assessment.

5.9 – Longwall Gate Entry Track Fire Risk Assessment Case Study

Mine I is a large underground longwall coal mine. The mine employs over 400 miners and operates three longwall and six continuous miner sections. The mine has track transportation throughout main headings and into development panels (*Figure 34*). The three-heading longwall development panel has one entry that has rails installed to the face area. This heading is also an intake for the fresh air ventilation to the working face and is the primary escapeway for miners. The belt entry contains the conveyor belt with a neutral split of air. The return entry contains the exhausted air from the working face and is this section's secondary escapeway. The major hazard evaluated at this site is a fire on the track entry of a longwall development panel, where smoke from the face travels to the working face obstructing egress through the primary escapeway.

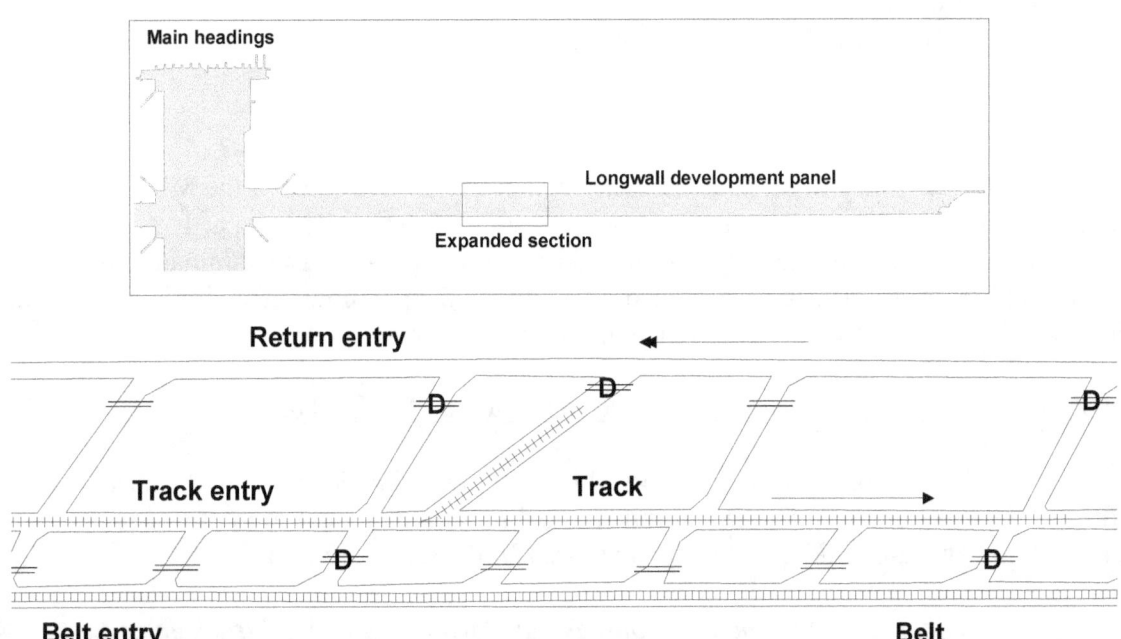

Figure 34 - Site conditions at Mine I showing the 3-entry development panel with direction of air flow.

5.9.1 - Risk Assessment Scope

The mine decided to review the risk related to fire hazards in the track entry of a longwall development panel considering the operation of relevant equipment and other variables. The operator did attend a NIOSH-sponsored training class and had expressed a desire to participate in the pilot project. The project objectives were scoped at the training session.

5.9.2 - The Risk Assessment Team

The team was made up of persons employed at Mine I as well as from the parent company and contained the following representatives from the workforce:
- Assistant mine superintendent
- Master mechanic
- Safety supervisor
- Fire prevention manager
- Corporate manager of fire prevention and mine rescue
- Motorman (labor)
- Miner operator (labor)
- Two subject matter experts
- NIOSH Observer
- Facilitator – MISHC (University of Queensland)

5.9.3 - Risk Assessment

The longwall track entry fire risk assessment did not use all five steps of the MHRA. A formal risk ranking of all potential unwanted events was not performed. This is partly due to the short time frame allotted for this activity and the desire to focus on a robust examination of prevention controls and recovery measures. It is also likely due to the potential difficulty in determining difference in the likelihood of occurrence of identified unwanted events.

5.9.3.1 - Step 1, Identify and Characterize Major Potential Mining Hazards

The first step in the risk assessment involved identifying and understanding the hazards related to a fire in the track heading. The team brainstormed the potential heat sources and fuel sources that might be available mid-panel to create a fire (*Table 43*).

Table 43 - Fuel and heat sources found within a longwall track entry.

Fuel sources	Heat sources
Materials on flat cars	Locomotive and mantrip motors
Some flammables on the locomotive, paint, coal dust, grease, oil, hoses, batteries, garbage	Locomotive Batteries
Stores in cross-cuts	High-voltage equipment, power cables
Coal	Welding and cutting
	Compressors
	Rock dusters

The team then decided to review risks related to a fire due to any source listed in *Table* 43 and located mid-panel within a track entry (intake air) in a development panel. It also agreed to the potential important characteristics listed in *Table 44*.

Table 44 - Important longwall track entry characteristics to be considered in the risk assessment.

	Track entry characteristics
1	Any variables or variations in development panel, e.g. dips
2	Equipment, primarily battery operated locomotives, portal buses (mantrips), rock dusters and compressors
3	Smoke conditions at the face would be dependent on the location of the fire, i.e. fire outby the face and near the main entry would fill both the fresh air intake and the track entry intakes with smoke, while a fire close to the face and far from the mains would only fill the track entry intake with smoke
4	Welding/cutting (hot work) activities sometimes occurred in the track entry intake
5	High-voltage cables were located in the track entry intake

5.9.3.2 - Step 2, Rank Potential Unwanted Events

After discussing and ensuring understanding of the above hazards, the team decided it did not have sufficient time or information to explore a risk ranking exercise with the WRAC tool. Instead, the team focused on identifying a list of ten consequences of a longwall track fire (*Table 45*).

Table 45 - List of acceptable and unacceptable consequences from a longwall track fire.

	Consequence	Risk rank
1	Loss of power to panel and face ventilation lost	Acceptable
2	Communication line is lost	
3	Roof fall in heading due to heat	
4	Discharge water line cut	
5	Compressed air line lost	
6	A small fire becomes a big fire	Unacceptable
7	Persons affected by smoke at face	
8	Person trapped by smoke	
9	Persons trapped or overcome (can't escape)	
10	Fire ignites gas in panel	

The team decided to combine the important characteristics of a longwall development track entry (*Table 44*) with the list of high-consequence unwanted events (*Table 45*) to rank the risk of a longwall track entry fire. The four highest risks are listed below:
1. Fire on a locomotive or portal bus (mantrips)
2. Electrical high-voltage fire
3. Welding or cutting fire (hot works)
4. Rock duster battery vehicle / compressor fire

5.9.3.3 - Step 3, Determine Important Existing Prevention Controls and Recovery Measures

The BTA method was used by the team to review and discuss the current controls in place to reduce risks related to the four high-consequence hypothetical fire events listed above. The complete BTA analysis is shown in Appendix A. This risk assessment compiled an extensive list of priority existing controls for event prevention and consequence minimization. Fifty-five existing key prevention controls (PC) and thirty-four existing key recovery measures are listed. Over 60% of the existing prevention controls were directed at the *fire on a locomotive or portal bus* potential unwanted event, demonstrating the mine's high concern with this risk.

An analysis of the 89 existing prevention controls and recovery measures provides a unique opportunity to examine the characteristics of the controls used by this mining operation. The character and effectiveness of controls was discussed under the topic of the hierarchy of effective controls (Section 4.4). The hazards unique to this risk assessment were associated with the use of a three-entry longwall gate entry design. The track entry was a designated fresh air entry. If a fire occurred in this entry, smoke would eventually make its way to the working face. The hazards associated with using track air to ventilate the working faces are the key component of this risk assessment. Therefore, if the need to use the track air current to ventilate the working face is eliminated then the hazard (EH) is eliminated. The most effective control, hazard elimination, was not discussed during this risk assessment.

All of the 89 controls fall in the other control categories: minimize hazards (MH), physical barriers (PB), warning devices (WD), procedures (P), and personnel skills and training (PST) (*Figure 35*). At this mining operation, there is a reliance on procedures (P) to mitigate the risks associated with the longwall gate entry track fires. Fifty-four percent of the existing prevention controls and recovery measures were classified as procedures (P). The rest of the controls were distributed, somewhat evenly, among the remaining categories. Most of the prevention controls that were categorized as minimizing the hazard (MH) are focused on the machines being designed and built to specifications that incorporate distinct safety features, i.e. enclosed compartments, fuses, breakers, de-energizing capabilities, etc. Also, more warning devices (WD) are used as recovery measures than prevention controls.

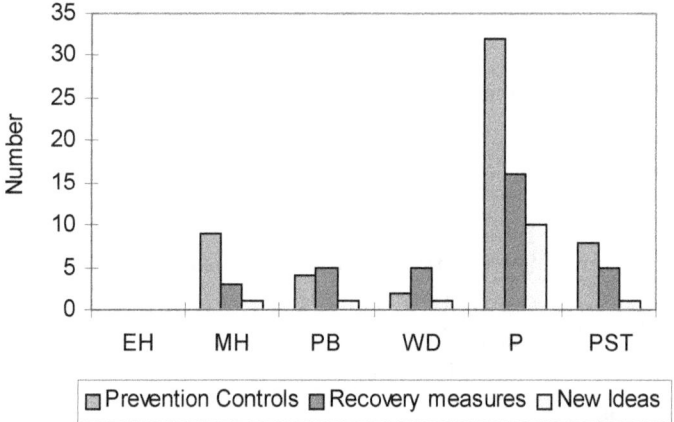

Figure 35 - Distribution of prevention controls and recovery measures for the longwall gate entry track fire risk assessment.

Many of the existing controls could be grouped by the issues they addressed. For example, the controls associated with the locomotives and portal buses can be grouped by design issues, maintenance issues and operational issues, as detailed below.

> *Locomotive and Portal Bus <u>Design Issues</u>:* The locomotives and portal buses are designed to standards with fuses, breakers and resistors. Many locomotive cables are protected in conduits. Radio-controlled communication is available on the locomotive to obtain assistance/advice about abnormal operations. Battery charging issues are minimized because the locomotives are always charging the batteries when in contact with a trolley wire in main headings. Also a gage indicating charge level is located in

operator compartments. Resistors are closed-in, blocking trash and other fuels from these heat sources. Fans are installed on some locomotives aiding in additional cooling. A red light indicator tells the operator when the brakes are on. Normal braking is electrical so overheating of the other brake is unlikely. All locos are fitted with heat sensors and manually initiated fixed fire suppression and hand-helds (20 lbs). Fire extinguishers are located within easy reach of the locomotive operator.

Locomotive and Portal Bus Maintenance Issues: All locomotive and portal buses are subjected to weekly maintenance checks. Mine personnel inspect new equipment and rebuilds before they are used. In addition, a state electrical inspector certifies that all major rebuilds are completed to standards. Battery maintenance includes cleaning, watering, checking for dead cells, etc. and all battery rebuilds are done to a mine specification. Fixed and hand-held fire equipment used on this equipment are checked regularly. Weekly inspections are completed to check for faults or discharge in the fire alarm system. Finally, a certified contractor does the maintenance inspection on fire suppression system every six months.

Locomotive and Portal Bus Operations Issues: Locomotive loads are always at least 10 ft from heat sources and only hydraulic oil and wood fuel sources are transported. Load guidelines are applied at the mine to avoid oversized/shifting loads that might derail the locomotive. Any explosives are hauled separate from all other supplies and transported in specialized containers. Operators receive special training and are required to perform pre-operation checks, i.e. test brakes, tram, check batteries and fire extinguishers. Safe Work Instruction and Best Practice teams sometimes observe pre-operations inspections. Operators are trained to open breakers and take plugs off batteries if there is a short, and if that does not work then the main lead is disconnected. Operators are aware of hot resistors and will stop operations and let resistors cool if overheated. Operators will smell for brake heat and look for abnormal operation, e.g. low power. Abnormal operation is reported to the supervisor and the maintenance shop. Supervisors know the capability of individual motor operators and they select competent operators for heavy/difficult loads. Operators are trained in the use of hand-held fire extinguishers every 2 years.

The rock duster and air compressor also had specific existing controls.

Rock dusters and air compressor issues: the Rock dusters and air compressor are designed to standard and inspected before underground use. Dedicated crews take care of charging and inspecting this equipment, including an operator being in the area during operation. There is a weekly electrical check of the equipment by the maintenance department. The rock dusters and air compressors have fixed, automatic and manually operated fire suppression systems directed at the battery areas. There are weekly inspections to see if the fire suppression system is faulted and if an alarm occurs, does it discharged. A certified contractor does the maintenance inspection on the fire suppression system every 6 months. All rock dusters and air compressors have a mounted hand-held fire extinguisher. Locomotive operators check the oil level in compressors.

Within the longwall gate entry environment, several other important issues are highlighted by the quality of the existing controls applied to them, as follows.

Track Issues: Rail maintenance program requires a given area of track to be inspected every shift. All tracks are installed and maintained to standards. The mine examiner examines the track during the pre-shift examination. Locomotive operators report any track issues to their supervisor.

High-Voltage Electrical Apparatus Issues: High-voltage cables are shielded and some are guarded and are located in rib/roof corner, reducing likelihood of damage. Circuit breakers/GFCI/pilot circuits are installed to protect the system form overload and fault fires. Monthly tests are done and recorded on these devices and on the jackets and insulation. Cabling is hung to regulatory requirements. The high-voltage system is designed to de-energize quickly (GFCI, etc.).

Housekeeping Issues: Each shift is responsible for cleaning up trash in their area. Trash is bagged and put on empty car. The locomotive operators pick up trash in outby areas around tracks.

Cutting and Welding: The mine has strict procedures for cutting and welding underground. Qualified persons must be present when cutting or welding to make methane gas checks and checking the area before they leave. Hand-held detectors are part of welding equipment used underground. Fire protection and rock dust is included in the welding and cutting procedure. Where possible a charged water line is also taken to the cutting or welding area.

Several important issues related to existing recovery measure were identified by the risk assessment team.

Gas Monitoring Issues: There is a real time continuously monitoring system that detects carbon monoxide (CO). Measurement points are located every 2500 feet in track heading. This system is linked to the underground bunker and outside surface hoist house where it is continuously monitored by a designated person. The system alarms at 5 ppm CO (alert) + 10 ppm CO (alarm). In addition, the system has a malfunction alarm. A designated person reacts to the alarms by 1) notifying shift foreman and other persons in affected area with both underground radios and telephones, and 2) checking CO detector. Also, there is a CO alarm at the conveyor tail (10 ppm CO) in the panel.

Fire Fighting Training Issues: Locomotive operators are trained in fire fighting every two years. Persons are trained that an air line can be charged to a two-inch water line to provide fire fighting water to the track heading. There is also a return water line that can supply water to fight a fire until air pressure to the face is lost. The mine has a designated Responsible Person (RP) who is notified when a fire occurs. All persons evacuate the mine if a big fire is identified. RP makes decisions about actions to be taken underground to fight fire, change ventilation, etc.

Emergency Egress Issues: Persons on face are trained to put on M20 self-rescuers, leave the panel if dense smoke is in the intake and meet at the power center, grab an extra SCSR, tag together, go to return, and use lifeline in return to egress the section. The M20 has a 20-minute supply of oxygen. There is a cache of 1-hour SCSRs at the load center and on mobile equipment, and caches are located 5700 feet in the intake track entry and 5700 feet in the return but staggered every 2850 feet down panel (staggered). Lifelines lead to caches and there are two cones on line to alert that a door or SCSR cache is present. There is a practice egress using the escapeways every quarter (alternating between the intake and return entries). Caches are located in cross-cuts with doors in stoppings. Persons are trained to take an extra SCSR. Per MSHA requirements barricading materials have been located in panels and persons have been made familiar with methods of building barricades. There are trained and qualified mines rescue teams available to attempt underground rescue. The Mine Emergency Response Plan includes external and internal communication, external medical services, family notification, security, etc.

5.9.3.4 - Step 4, Identify New Prevention Controls and Recovery Measures

Fourteen new ideas were identified by the team during the risk assessment to further reduce the longwall track fire risks at the mine (*Table 46*). The particular BTA that was responsible for each new idea is provided in Appendix A (*Table 56* and *Table 57*). Seven new ideas address prevention control measures and seven recovery measures. The hierarchy control categories for the new ideas are also dominated by procedure (P) controls (*Figure 35*).

Table 46 - New prevention control and recovery measure ideas for the longwall track fire event organized by category. NOTE that the new idea numbers (NI) correspond to the new ideas listed in the BTA for the risk assessment (Table 56 and Table 57).

Design	NI7	Investigate changing or modifying loco resistors to perform under load without overheating (MH)
	NI9	Investigate whether fixed fire suppression can be located over/at compressor (PB)
	NI10	Put one joint of fire hose on loco to be carried on the track jeep at all times (P)
Maintenance	NI1	Reinforce and follow the requirements of the maintenance program for batteries, consider checklists/verification that it is being followed (P)
	NI3	Add checking gauge accuracy in the battery maintenance program (WD)
Operations	NI2	Investigate defining a specific percentage battery charge that is minimum to enter panel (P)
	NI4	Investigate whether there is an identifiable level of complexity/experience for major loads, thereby creating a list of heavy load operators (P)
	NI5	Reinforce the need, during pre-operation inspection, to remove any baking soda that has been used to absorb water on batteries so that it doesn't become a conductor (P)
	NI6	Add checking inside the loco resistor area (lift lid) for any combustibles to pre-operation inspections (NOTE that dust can get into resister compartment) (P)
	NI8	Make operators aware that, if possible, when there is a small fire or smoke from a loco/mantrip/rock duster there may be an opportunity to reduce/stop smoke to the face by putting equipment into a switch/spur track and open man-door into the return heading to short circuit into the return (P)

Fire and Emergency Response	NI11	Add clarification to ER training, re: egress in light smoke – i.e. if light smoke in intake use transportation to exit as far as possible* then cross to return to egress [* smoke is too dense to see ahead] (P)
	NI12	Make sure the caches are located in cross-cuts with doors in stoppings (P)
	NI13	Reinforce the need to put self-rescuer or, if closer by, don SCSR as soon as any smoke is detected (issue: may get worse and easier/safer to don the SCSR now) (PST)
	NI14	A method should be developed to access stopping doors at caches to check if intake is fresh air so that a person can remain attached to lifeline and/or team. The method should be included in 90-day ER training. (P)

NI – New Ideas
EH – Eliminate Hazard
MH – Minimize Hazard
PB – Physical Barrier
WD – Warning Devices
P – Procedures
PST – Personnel Skills and Training

5.9.3.5 - Step 5, Discuss Implementation, Monitoring and Auditing Issues

Assuming that the information provided in the risk assessment is accurate, an increased focus on monitoring and auditing of the key identified controls would appear to provide an opportunity to effectively reduce the risk of fatalities related to underground fire in the track entry at Mine I. The risk assessment team identified 89 existing controls and 14 new ideas. Procedures dominate both the existing and new prevention controls and recovery measures for this mine site. Procedures are known to have a potential for human error. This requires a thorough examination and audit effort. This mining operation will address these needs through a Safe Work Instruction program and Best Practice teams that periodically observe the quality of many existing controls.

At the end of the risk assessment, the 14 ideas for new potential controls and recovery measures were submitted to mine management in the form of an Action Plan (Appendix B). The Action Plan lists each new idea and contains additional columns to identify who will investigate the idea, when the investigation will be completed, and what specific action will be required.

5.10 – Change of Mining Method Risk Assessment Case Study

A risk assessment was performed at an underground metal mine (Mine J) to investigate major hazard potentials associated with the management of change. Mine J operates in a steep, near vertical, ore body with the captive (raise access) cut-and-fill stoping method (*Figure 36*). In this method, the ore is mined by successive flat slices, working upward. After each slice is blasted down, all broken ore is removed, and the stope is filled with waste up to within a few feet of the back (roof) before the next slice is taken out. The term captive implies that access to the stope is solely through vertical access raises that are confined to that stope. During production, one of these raises can contain ore. This method requires three miners per stope.

The scope of the risk assessment was limited to those events that would have the potential to fatally injure the miners working in the captive stopes. The primary change related to the move from mechanized (drift access) cut-and-fill to captive stoping is considered to be 1) the occasional limitations on escape from the work area to a single ladder-way in the access raise, and 2) the occasional requirement of miners to work under unsupported brows. Also, the captive cut-and-fill stoping method relies more on miner hand-work and less on mechanized equipment than the previous mining method. Members of the mine staff attended a NIOSH-sponsored MHRA training course and expressed an interest in participating in the study.

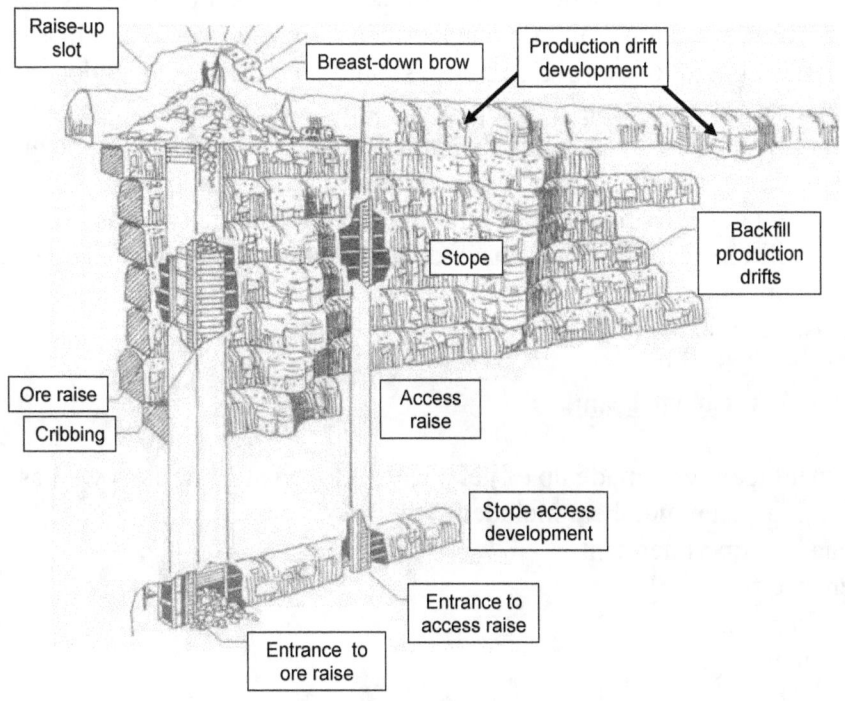

Figure 36 - Diagram of Captive Cut-and-Fill mining method.

The captive cut-and-fill stoping method is a complex practice that has been widely used around the world to skillfully respond to changing conditions within the ore body. The widths of the mining space can shrink and expand in concert with the ore thickness. In the last few decades some mining operations have used a more mechanized stoping method with tire-mounted jumbo-boom style face drills to blast the mine opening and load-haul-dump (LHD) vehicles to efficiently move the muck (broken ore) out of the production stope. The design of the stope is highly dependent on the size and maneuverability of the mobile equipment. When the ore body falls below a certain thickness, mechanized stoping can become inefficient. This is the case with Mine J.

While the captive cut-and-fill stoping method lacks large mobile mechanized equipment, there is still a strong reliance on a wide range of mining equipment. Every stope has a complement of electric slushers[5], air tugger[6], jackleg drill and a mucker.[7] There is also a wide range of electric fans, tools, lights, and phones that require a transformer and hundreds of feet of wire. Compressed air and water are also brought into the stope. In addition, each stope is serviced by diesel haul trucks and tractors and an explosives magazine.

5.10.1 - Risk Assessment Scope

The scope of this risk assessment is to identify the major hazards and risk potential associated with the change of mining method to captive cut-and-fill stoping and to evaluate existing prevention controls and recovery measures for adequacy in controlling identified risks. A change of mining method risk assessment represents the most complex MHRA because it can consist of a number of smaller related risk assessments. Each hazard examined identified existing prevention and response actions that were considered as important to maintain. Further discussions of what other actions might be undertaken to further reduce the likelihood of the subject event occurring and to improve response should it occur were also documented for management review of the concepts developed. In some cases a work process flow chart of the specific portion of the mining cycle being examined was developed so that the group could consider where prevention and early response actions could best be placed. A significant amount of time must be spent considering the new mining process, and the composition of the risk assessment team will change as different hazards require specialized knowledge bases. The output is information to assist Mine J in the development of the approach to captive cut-and-fill mining so that risks are managed to a level that is acceptable.

5.10.2 - The Risk Assessment Team

The risk assessment team was made up of persons employed at Mine J, as well as from its parent company. The initial team members included:
    Maintenance superintendent
    Maintenance foreman
    General foreman
    Shift foreman

---

[5] A <u>slusher</u> is a blade or bucket that drags the broken ore within the production drift to the dump point at the top of the ore raise.

[6] A <u>tugger</u> is a small, semi-portable hoist, powered by compressed air or electricity, to raise supplies and equipment within the access raise.

[7] A <u>mucker</u> is the device that loads the broken rock out of the ore raise and into haul trucks for transportation out of the mine.

Two captive stope miners
Rock mechanics engineer
Safety coordinator
Director of safety
NIOSH matter experts
NIOSH observer
Facilitator – MISHC (University of Queensland)

The risk assessment team composition changed as the team focused on additional hazards or different work processes.

5.10.3 - Risk Assessment

The structure of the risk assessment methods used was a multiple layered approach beginning with a semi-quantitative risk ranking exercise, using a WRAC, to prioritize hazards for further examination. The eight highest priority risks were then examined in more detail utilizing the BTA and work process flow chart methods. This risk assessment required that any hazard suspected to be associated with the new mining method needed to be evaluated.

5.10.3.1 - Step 1, Identify and Characterize Major Potential Mining Hazards

For a change of mining method risk assessment, there is a potential for numerous major hazards. Some of these hazards exist within the current mining method, others are specific to the new mining method and, as such, may be new to the mine. In this case, the hazards can only be identified and characterized after the mining process has been segmented into distinct phases. This first operation was completed by the first team over the course of one day where the captive cut-and-fill stoping method was broken down into 11 phases with an internal loop for the repetitive portions of the cycle (Phase 6 to 10).

1. Stope access development
2. Vertical raise preparation (two raises per stope)
3. Vertical raise-up slot (locally referred to as a beanhole)
   a. Jackleg raise mining
   b. Longhole raise
4. Preparation for ore production by lining the vertical raise with wood cribbing
5. Production drift development
   a. First connecting vertical raises (I-drift)
   b. Develop drift to its full horizontal length of ~ 200 ft (sill drift)
6. Raise-up a slot in the production drift
7. Place and decant sand fill in production drift, and relocate equipment
8. Breast-down the brow along the production drift
9. Extend cribbing in vertical raises in preparation of placing sand fill in the production drift
10. Reset equipment for next raise-up of the production drift [return to Phase 6 to repeat 6 to 10 for 200 ft vertically (~20 cuts)]
11. Remove equipment from stope.

Next the team listed the types of related hazards that should be considered (*Table 47*).

Table 47 - Hazards associated with captive cut-and-fill mining.

	Potential hazards	Where	When
1	Ground fall	Production drift	Production from drift, raise-up slot, breast-down
2	Inrush of previously place sand fill	Production drift, access raise, stope access drift	Placing and decanting sand fill and relocating equipment
3	Electrocution	Stope and stope access drift	Energizing, splicing, moving, etc. electric wires
4	Air pressure	Stope access drift, access raise and production drift	Connecting air lines, pumping sand fill and operating drills
5	Water pressure	Stope access drift, access raise and production drift	Connecting water lines, pumping sand fill and operating drills
6	Diesel / hydraulics	Stope access drift	Hydraulic failure leads to fire on diesel equipment
7	Explosives	Production drift, breast-down, raise-up slot, dynamite magazine, transporting in access raise	Transportation, placing in blastholes, pre-detonation and explosives that failed to detonate.
8	Falls (drawpoints gravity)	Ore raise and access raise openings	When moving through or approaching access or ore raises
9	Equipment temperature	Stope and stope access drift	Overheating of diesel engines or electric motors
10	Slusher or tugger cable tension	Production drift	Overstressed or over worn cables used during slusher or tugger activities
11	Slusher setup and anchor	Production drift	Production from drift, raise-up slot, breast-down
12	Mechanical energy of equipment	Stope and stope access drift	When operation or preparing to operate diesel, electric, hydraulic or compressed air equipment
13	Dangerous gasses	Raise-up slot, production drift	Formation gases released during drilling or blasting or dangerous gases associated with diesel particulate, blasting or fires
14	Slip / trip	Stope and stope access drift	Moving over uneven and rocky surfaces or tripping over equipment

Once the mining phases and hazards were identified, the team was ready to apply risk analysis methods.

5.10.3.2 - Step 2, Rank Potential Unwanted Events

The team considered each mining phase and each hazard individually in order to systematically identify potential unwanted events. The WRAC produces information about the mining phases and specific unwanted events. Eighty-three potential unwanted events were identified (*Table 48*). Every mining phase contained at least one and most had many events. The number and significance of these events attest to both the complexity of the problem and the collective knowledge of the team.

Table 48 - Priority listing of potential unwanted events associated with phases in the captive cut-and-fill stoping method at Mine J.

Mining phases	Potential Unwanted Event
Stope access development (1)	Methane due to drilling into pocket
	Cable bolts come out before grouting
	Loss of long hole steel causes impalement
	Drilling into a miss hole with explosives causes explosion
	Major fall of ground while bolting
	Blast into diamond drill hole
	Manual handling of electric cables
	Mucking into a misfire causes explosion
	Fall of ground between supports
	Equipment fire
	Crushed by mobile equipment
	Explosion occurs from manually handling explosives
Raise preparation (2)	Improper location under roof causes fall
Initial raise development - jackleg (3a)	Drill into water / gas source causes falls
	Miss hole hit by drilling in bean hole causes explosion
	Rockfall in bean hole w / jackleg
Initial raise development - long hole (3b)	Longhole breaks through into lower level where people are working
	Longhole blasting breaks through into where people are working
Lining the vertical raises with wood cribbing (4)	Welding chutes (truck)
	Electrocution occurs when welding due to water exposure
	Rigging failure causes release of tension
	Timber failure occurs
	Person crushed against ribs when positioning equipment
	Person with less than adequate familiarity accesses stope area
	Person hit by material falling down skip shaft
	Fire in the intake sends smoke into stope
Production drift development - I drift (5a)	Miss holes in I drift explode when drilling
	Electrical shock due to damage during blasting
	Gasses encountered re-entering area after blast
	Fall of ground during first couple rounds
	Person falls into chute when mining nearby
	Person knocked back into chute by blow pipe
	Equipment fire in the intake while person is in I drift / sill
	Person hit by rock fall while accessing unsupported ground to set up slusher
	Persons not familiar with jacklegs operate that equipment
	Person falls into holes due to incorrect covers
Production drift development - sill drift (5b)	Electrical fault occurs when plugged into wrong power source
	Person hit by parts of mucker if hits rib
	Equipment fire at face
	Protruding ground support when operating mucker

Raise-up slot (6)	Hang-ups cause cave-in or drill injuries
	Rock bursts (strain bursting) during raise development
	Gasses cause exposure to bad air after access to raise-up
	Water air pressure release hits person
	Loss of floor (muck / staging) pulls person in
	Loss of floor occurs in new raise-up due bulkhead or mucking out
	poor location of eye leads allows equipment to fall in raise
	Fall from height
	Person injured due to skip problems having impact on man access
	Miss holes after raise-up causes unsupported ground
	Persons exposed to unsupported ground in crossing over muck pile
	Damage to ground support while slabbing
	Person hit by roof fall when trying to support raise-up
Place and decant sand fill (7)	Failure to decant causes crib failure later in mining
	Person gets stuck, sinks or asphyxiates while accessing fill area to repair line
	Release of pressure when line plugs
	Miner filling sand fill falls into raise when cribbing fails
	Decant into ore pass causes muck to blow out
	Person working alone when sand filling
	Decant water causes failure in another area as it runs out of stope
	Person falls off timber while filling down manway or into fill
Lining the raise-up with wood cribbing (8)	Loss of control of timbers causes crushing
	Person falls into chute
	Person hit by timber when dropped down man access / bean hole
	Rigging failure occurs when relocating equipment and persons hit
	Person falls off timber >10'
	Person falls into beanholes when installing timber sets
Breast-down the brow (9)	Early initiation of blast occurs when person in manway due to safety fuse
	Brow rounds blows rock onto manway
	Unexpected geologic structure causes ground problem
	Offset / jogs in stope causes stresses and other problems
	Mistimed blast in reef causes impact in other area of mine through to another level
	People in other level when mining blasts / accesses area
	Survey error leads to mining out in another level not as planned
	Hanging walls burst / slab out
	High stress occurs in pillars affecting stope ground / access
	Access sand filled stope and collapse occurs
	Inadequate crown pillar size leads to collapse when another level accessed
Reset equipment for next stope (10)	Electric shock occurs in set equipment when setting up
	Equipment not set in right location and / or set insecurely
	Persons injured relocating slusher under its own power (not hooked up correctly)
Remove equipment from stope (11)	Cable failure lowering equipment causes accident
	Failure of chain fall causes persons to be hit by equipment

Once all phases of captive cut-and-fill stoping are considered, each unwanted event is risk ranked using the cooperating company's risk matrix (*Table 49*).

Table 49 - Risk Matrix used by cooperating mining company.

		Likelihood of Occurrence				
		1 – Highly unlikely	2 – Not expected	3 – Slight potential	4- Moderate potential	5 – Highly likely
Consequence	1 - Immaterial (I)	I-1	I-2	I-3	I-4	I-5
	2 - Low consequence (LC)	LC-1	LC-2	LC-3	LC-4	LC-5
	3 - Moderate consequence (MC)	MC-1	MC-2	MC-3	MC-4	MC-5
	4 – High consequence (HC)	HC-1	HC-2	HC-3	HC-4	HC-5
	5 - Disaster (D)	D-1	D-2	D-3	D-4	D-5

The risks are calculated by the product of the *likelihood* times the *consequence,* and they range from a high of 20 to a low of 3. The highest priority risks identified by the WRAC process are listed in *Table 50*.

Table 50 - Highest ranked risks potential unwanted events associated with captive cut-and-fill stoping.

Potential unwanted event	L	C	R
Explosion occurs manually handling explosives	5	HC	20
Person hit by roof fall when trying to support raise-up	5	HC	20
Crushed by mobile equipment	4	HC	16
Person falls into holes due to incorrect covers	4	HC	16
Person falls into beanholes when installing timber sets	4	HC	16
Person falls off timber while filling down manway or into fill	4	HC	16
Fall of ground between supports	5	MC	15
Equipment fire	3	D	15
Damage to ground support while slabbing	5	MC	15
Improper location under roof causes fall	3	D	15
Rockfall in beanhole w / jackleg	5	MC	15
Person with less than adequate familiarity accesses stope area	5	MC	15
Equipment fire in the intake while person is in I drift / sill	3	D	15
Person hit by rock fall while accessing unsupported ground to set up slusher	5	MC	15
Persons not familiar with jacklegs operate that equipment	5	MC	15
Equipment fire at face	3	D	15
Mucker operator hit by protruding ground support	5	MC	15
Person hit by material falling down skip shaft	5	MC	15
Fire in the intake sends smoke into stope	3	D	15
Decant water causes failure in another area as it runs out of stope	5	MC	15
Hanging walls burst / slab out	3	D	15
High stress occurs in pillars affecting stope ground / access	3	D	15
Access sand filled stope and collapse occurs	3	D	15
Inadequate crown pillar size leads to collapse when another level is accessed	3	D	15

L = Likelihood
C = Consequence
R = Ranking

Many of these events had similarities and the facilitator recognized that the list of events, requiring detailed analysis, needed to be reduced. Therefore, the group agreed to use maximum likely consequence and set the highest risk level at a multiple fatality potential (consequence = disaster, *Table 49* and *Table 50*). A decision was then made to group the highest ranked potential unwanted events into smaller, more generalized, events. *Table 51* lists the four highest priority risks identified by the team that could result in a multiple-fatality event.

Table 51 - Highest priority risks capable of producing a multiple-fatality event.

Detailed analysis using the BTA	
1	Equipment fire in intake airway with persons in the stope at the face
Detailed analysis using the work process flow chart	
2	Unfavorable location of the footwall lateral drift causes unstable ground condition in the access drifts
3	Stope geometry is poorly defined by diamond drilling causing mining into sand fill drifts
4	High stress conditions are poorly defined by geotechnical modeling causing rock bursts

5.10.3.3 - Step 3, Determine Important Existing Prevention Controls and Recovery Measures

One of the risks was analyzed with the BTA and three with a work process flow chart. The process took about 5 hours for a large, diverse group after 2 hrs of training in the principles of risk assessment and management. Through these approaches, the team identified an extensive list of existing prevention controls and recovery measures. Most of the controls listed represent the mines internal Best Practices. A common theme of discussions of these controls was the dependence on a few individuals to ensure compliance, with no formal auditing methods for some critical controls.

Potential Unwanted Event 1 - Equipment fire in intake airway with persons in the stope at the face: The team identified the locomotive in the intake drift of the mine and haul trucks operating in the stope area and muckers operating in the stope area as the three primary sources of a fire with the potential for major consequences. The team identified 49 existing equipment fire prevention controls (*Table 52*). Thirty-one apply to all existing equipment, 12 are specific to locomotives, and 6 pertain to haul trucks and muckers. A total of 8 hours was spent examining the equipment fire hazards and controls.

Table 52 - Priority existing prevention controls and recovery measures for equipment fires in the stope and stope access drift.

	PC1	Scheduled preventative maintenance on equipment is done every 250 hours, including hose inspection and change out (P)
	PC2	All work is done by mechanics (P)
	PC3	Hoses are four-braid, higher standard on locomotives, haul trucks and muckers (PB)
	PC4	Hose routing issues, such as damage that is found in inspection or maintenance, should lead to rerouting of hoses to correct the problem. Corrections should occur such as relocation, shielding, guarding, etc. (PB)
General equipment fire	PC5	The supply fuel lines are hard over the engine area, secured and located in a low location so any minor fuel leaks will not drip on hot surfaces (MH)
	PC6	Any return fuel lines are soft but four-braid hoses are used and located away from heat sources (MH)
	PC7	Heat wraps are located on hot surfaces of locomotives, haul trucks and muckers (PB)
	PC8	Wires are in looms to keep them in place and protect them from damage (PB)
	PC9	The electrics are maintained by trained personnel (2 levels) (PST)
	PC10	Locomotives, haul trucks and muckers should have circuit breakers (MH)
	PC11	New locomotives, all haul trucks and muckers should have wet brakes (MH)
	PC12	Mobile equipment operators are trained to the SOPs and other levels of operator are trained too (PST)

	PC13	Daily operator inspections are made on locomotives, haul trucks and muckers to identify damage, leaks, flammable materials, etc. Operators are trained in the use of a Yellow Card for inspection. (P)
	PC14	Vehicles should have hydraulic pressure and temperature indicators that tell the operator about operating conditions and abnormalities (WD)
	PC15	Vehicles should shut down if they are overheated (MH)
	PC16	Preventative maintenance should include a power wash (P)
	PC17	Speed is controlled on all equipment by a hard barrier blocking the use of 4th gear (MH)
	RM1	Fire suppression systems are on all locomotives, haul trucks and muckers, activated by operator in the cab (MH)
	RM2	5-lb hand-held fire extinguishers are located in equipment cabs, with operators trained annually (P)
	RM3	A fire in the intake should require immediate evacuation of the mine (P)
	RM4	Fire emergency procedures should cover dispatch activating a computerized warning system (stench), using the radio and phone system to warn miners in the stopes about the fire (WD)
	RM5	All persons are trained in fire emergency procedures. Persons are trained to, on receipt of a fire warning, 1. Get out of the mine 2. If not possible go to refuge and stay in refuge until released, and 3. If cannot access refuge then barricade. (PST)
	RM6	All persons should have a CO chemical self-rescuer that can operate for an hour at 1% CO to help facilitate escape or access to the refuge (PB)
	RM7	There is a refuge chamber installed within 10 to 15 minutes travel from the stope, marked well (including air supply, communications, water, etc. for several people) (PB)
	RM8	Any required barricading in the stope is done with material and compressed air supply is available if not damaged (PB)
	RM9	Water supply in the stope is used to provide some protection in a fire (MH)
	RM10	There is a reliable (24/7) air compressor operating on the surface that can replace underground compressed air, activated by a manual control on the surface (MH)
	RM11	There is a brass-in and brass-out system to ensure that persons underground are accounted for (P)
	RM12	The mine has a looped leaky feeder system designed so that if it has a break due to a fire it still works (MH)
	RM13	There is a trained mines rescue team (MH)
	RM14	There is an ambulance and paramedics available in the surrounding area (MH)
Locomotives	PC18	The locomotive should shut down if it overheats (MH)
	PC19	Pressure failure should lead to total brake application on the locomotives (MH)
	PC20	Hydrostatic braking should provide an opportunity to use retard braking as an alternative to mechanical braking (MH)
	PC21	The locomotive, haul trucks and muckers have an air pressure gauge that tells the operator about operating conditions and abnormalities (WD)
	PC22	The locomotive is fuelled only on the surface unless there is a breakdown / fuel problem underground (MH)
	PC23	Engines are shielded from "blow in" materials / debris (PB)
	PC24	The fuel rail car is double walled (PB)
	RM15	There is fire suppression on the fuel rail car (MH)
	PC25	The cab should always separate the fuel rail car from the diesel engine (P)
	PC26	Special operational controls are in place when fuel is transported into the mine, including two specialized locomotive operators (P)
	PC27	There is a master switch on the locomotive to shut it off in an emergency but it is not easy to access (MH)
	RM16	Vehicle operators are trained to communicate a fire problem to the surface using a personal radio (PST)

	PC28	Supervisors should check two pieces of equipment per shift to ensure daily operator inspections have been done (P)
Haul trucks or muckers	PC29	Fuelling is done underground with quick disconnects for fuelling to decrease risk of vehicles driving away with hose in fuelling location (MH)
	PC30	Back pressure is indicated in the cab to tell the operator when the DPM is blocked by indicating pressure in the "red" sector of the gauge. The operator is trained to call maintenance in this situation (PST)
	PC31	A hot work[8] system is in place (P)
	PC32	There is a phone near where the two-yard mucker operates so that absence on the mucker is acceptable (WD)
	As above: see PC35 & RM43	

PC – Prevention Controls
RM – Recovery Measures
EH – Eliminate Hazard
MH – Minimize Hazard
PB – Physical Barrier
WD – Warning Devices
P – Procedures
PST – Personnel Skills and Training

Potential Unwanted Events 2, 3 and 4 – Major ground failure due to inadequate mine design with crew working in the stope trapped or fatally injured: Three of the high-risk events (numbers 2, 3 and 4, *Table 51*) were determined to be primarily related to failures in the stope design process. At this mine, a stope proposal is produced for every proposed stope prior to mining. The stope proposal contains detailed mining and operational information. The team decided to evaluate the stope design process. The team also determined that it needed to be reorganized to contain more staff with mine planning experience.

The existing stope planning process was mapped in detail, noting the points where decisions affecting the high-risk events occur and promoting an orderly discussion of the hazards. The primary output of the exercise was to recommend that a new step (*Figure 37*, step 3B) be inserted in the planning process to evaluate geo-mechanical issues in the Long Term Planning process (LTP). In addition, points of failure in the execution of the existing process steps were examined and, when the impact of failure was an increased risk of the subject event occurring, solutions and means of monitoring for compliance were developed as potential new controls.

---

[8] Hot works = welding, cutting, grinding, etc.

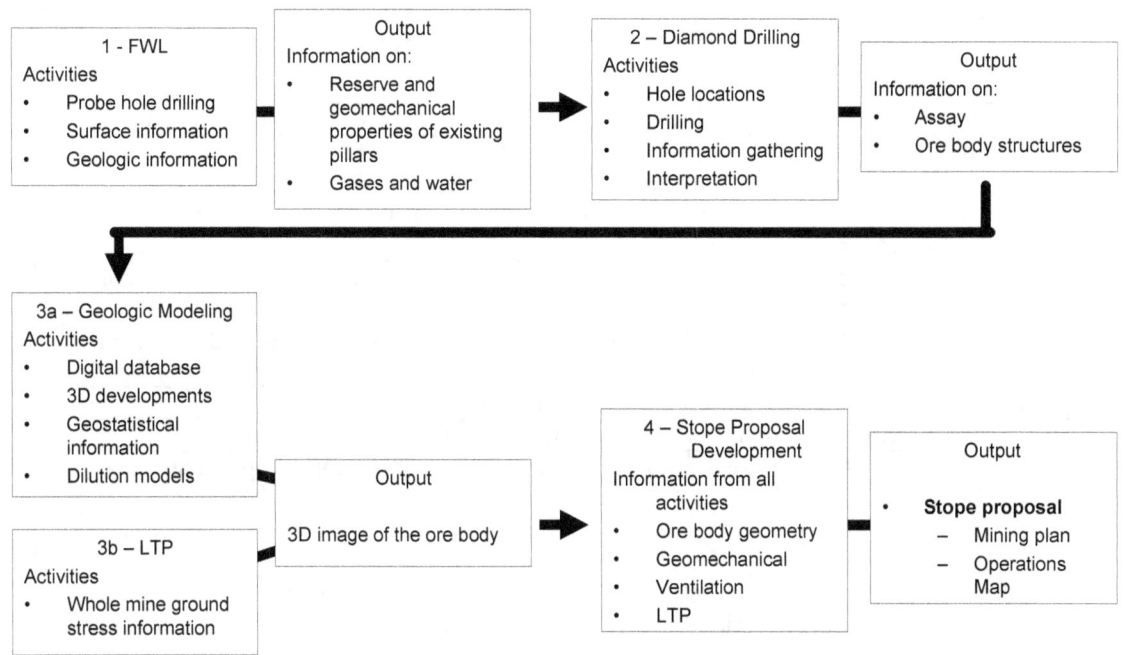

Figure 37 – A flow chart of the basic stope proposal and mine planning process.

5.10.3.4 - Step 4, Identify New Prevention Controls and Recovery Measures

Potential Unwanted Event 1 - Equipment fire in intake airway with persons in the stope at the face: After considering these existing controls the team identified additional steps to both prevent a fire from occurring and increase the likelihood that miners in the captive stopes will survive. These additional controls suggested by the team are identified in the following list (*Table 53*), again as either general or equipment-specific precautions.

Table 53 - New prevention control and recovery measure ideas for the equipment fire in the intake drift event.

	NI1	Document the locomotive, haul truck and mucker hose specification (four-braid) and locations standards so they continue to be applied consistently (P)
	NI2	Check to ensure that that the improvements on hose routing on all mobile equipment and other modifications resulting from found damage are gathered to apply to other equipment when they are maintained (P)
	NI3	Add a final inspection by maintenance after preventative or other maintenance to ensure engine / brake area is clear of debris (P)
Equipment fire in the intake entry	NI4	Consider the use of a heat gun to check the operating temperature of the brakes (P)
	NI5	Document the locomotive, haul truck and mucker fuel hose specifications (hard and soft) and locations standards so they continue to be applied (P)
	NI6	Check to make sure the positive lead is protected with a fuse and master switch (P)
	NI7	Document the locomotive, haul truck and mucker electrical cable / wire specifications and locations standards so they continue to be applied (P)
	NI8	Ensure that work done on under 25V, and related training, does not compromise fire exposure (P)
	NI9	Add a final inspection by maintenance after preventative or other maintenance to ensure engine / brake area is clear of debris (P)
	NI10	Check Diesel Particulate Matter (DPM) filters to clearly identify design and maintenance requirements (P)

	NI11	A hot work system should be set up at the mine (P)
	NI12	Review the welding / cutting SOP at other mines and apply to this mine (audit current situation versus new SOP) (P)
	NI13	Define the required welding competency and train / "ticket" welders / cutters (PST)
	NI14	Make sure all persons have been appropriately fire trained (PST)
	NI15	Test whether the refuges can be found in a smoky situation and consider ideas such as lanyards and lasers to help miners find the refuge (note that muck bays may be inadvertently accessed too) (PST)
	NI16	Put CO shut off, PED shut off, or an E stop in the stope to shut down the section ventilation fan in a fire situation (WD)
	NI17	Investigate automation of surface compressor supply to underground should underground compressor fail or be compromised (MH)
	NI18	Investigate ways to effectively barricade in the stope (PST)
	NI19	Investigate ways to supply more air into the stope (MH)
	NI20	Investigate a second top egress to improve survivability if fire traps people in the stope (MH)
	NI21	Add the use of stope water supply for fire events as a part of training for dealing with a fire if trapped in the stope to Emergency Training (PST)
Locomotives	NI22	Make sure inspection checks are done on locomotives as well as rubber tired equipment (P)
	NI23	Reinforce the need to review the risks related to underground fuel bays located in intakes entries (PST)
	NI24	Check to ensure that there is automatic fire suppression on all locomotives (P)
Haul trucks or muckers	NI25	Investigate the application of Wiggins fuelling hardware in underground fuelling locations (MH)
	NI26	Reinforce the need to shut down muckers when the operator is not on the machine (P)
	NI27	Investigate the ability to isolate a vehicle fire in the lateral footwall area (MH)
	NI28	Vehicle operators in the lateral footwall / stope access area should be made familiar with the refuge location and related actions required during a fire (where to go to barricade) (PST)

NI – New Ideas
EH – Eliminate Hazard
MH – Minimize Hazard
PB – Physical Barrier
WD – Warning Devices
P – Procedures
PST – Personnel Skills and Training

It is worth noting that most of the 25 new control measures are either directed to improve response to a fire to mitigate consequences or they increase the likelihood that existing good practices are followed uniformly through more rigorous audit systems and documenting of practices. Five of the new controls concern improving the ability of miners to secure refuge from smoke and maintain a reliable supply of emergency air. While the mine has refuge chambers and well-developed emergency plans, this exercise was the first detailed examination by the miners as to how they would secure a refuge from smoke should they find themselves trapped within the stope. Part of this detailed examination was a hard look at the vulnerability of the existing measures to damage from the event and formulation of a "plan B." The value of the process was achieved by having the miners mentally work through what steps they could take to react to problems. Innovative solutions were suggested and the gaps in protections needing further effort by mine staff were identified.

Also of note is the relative rank of the controls with regard to the typical hierarchy. Existing controls are a mix of engineering, 41%, administrative, 57% and monitoring, 2%, with none directed at eliminating the hazard of combustible fuel and oils due to the mining method's dependence on diesel equipment. In the new controls the mix is engineering, 24%, administrative, 52%, and monitoring, 24%. One new response control is directed at eliminating the primary hazard of a single egress from the stope by suggesting a second egress out the top of the stope to an independent airway.

Potential Unwanted Events 2, 3 and 4 – Major ground failure due to inadequate mine design with crew working in the stope trapped or fatally injured: The left-half of the BTA was performed, concentrating on modifications to the mine planning and design process to prevent an occurrence of a major ground failure. The consequences side of these high-hazard events was not examined due to time constraints and the low likelihood that response actions would alter the outcome of these rapidly developing catastrophic events. Twenty-four new prevention controls were identified (*Table 54*) that spanned the four distinct planning phases identified in *Figure 37*.

Table 54 - New prevention control ideas for stope design and mine planning.

1 – Improving Mine Planning Footwall Lateral Locations	NI29	Make Long Range Planning (LRP) more systems oriented, i.e. ventilation, utilities, mine method, service life, egress, etc. (MH)
	NI30	Purchase additional survey equipment (P)
	NI31	Audit that survey is being done (P)
	NI32	Communicate that job isn't done until surveyed (P)
2 – Definition by Diamond Drilling (DD)	NI33	Draft formal policy with exception requirements (P)
	NI34	Support concept that more development and probing provides more information and flexibility to this stage and increases chance of success (P)
	NI35	Receive timely Vulcan input in order for geologists to see changes in 3D model (P)
	NI36	Cross-train and temp hire for high logging demand periods (PST)
	NI37	Log info at hole when drilling bad ground (P)
	NI38	Increase amount of information collected from directional drill core near ore zone (P)
	NI39	Audit that layout design is followed (P)
3a – Geo Modeling	NI40	Improve retention of technical expertise including mine site experience (P)
	NI41	Investigate use of survey-based volume reconciliation (P)
	NI42	Continue focus on reconciliation process between face and mill (P)
3b – Stress Control Planning (new in the model)	NI43	Ground stress considerations need to be incorporated into long range and life of mine planning as well as stope proposal (P)
	NI44	Investigate and apply methods that gather useful ground stress information during mining for mapping (P)
	NI45	Gather stress info from instrumentation and map to aid planning (P)
	NI46	Look at other similar mining operations re: overall ground stress potentials (P)
4 – Stope Proposal Development	NI47	Consider stope proposal and face mine planning system at other operations to help develop their approach (P)
	NI48	Mentor new miners after stope school by placing with experienced miners in stope to learn plan (PST)
	NI49	Develop a standard that requires mining to minimum width to ore before widening for raise access (P)
	NI50	Establish expert captive stope prep and development crew (PST)
	NI51	Use 3D design images to introduce production people to the new mine method and area (PST)
	NI52	Use input from production to walk through 3D info and develop final plan with detail by step of mining process (P)
	NI53	Reinforce importance of scheduling services and equipment (P)

The 24 newly identified controls are made up of administrative, 64%, engineering, 24%, and monitoring, 12%. The newly added step in the planning process emphasizes engineering controls. The end product of the proposed changes would be to put in place a mechanism to remove the potential for exposure to the high-risk hazards.

5.10.3.5 - Step 5, Discuss Implementation, Monitoring and Auditing Issues

The overall acceptance and understanding of the risk assessment process by the teams at this mine was very good. The teams selected were energetic and knowledgeable in the subjects. Communication between team members during the process was thorough and members actively challenged each other's paradigms. All of these factors had a positive effect on the outcome of the process which developed a large number of suggestions for improvement embraced by management.

One issue encountered was the amount of time necessary to educate team members on the process. In this group, none had any exposure to formal risk management practices. As a result a half-day was required to train the team members. As risk management is applied to more areas of the mine operations, a significant training burden will occur.

A second issue encountered was the tendency of the team to try to solve every issue that arose rather than identify issues and move on. A skilled facilitator was required to keep the team on task. The facilitator was also challenged with recognizing team members who had difficulty expressing their thoughts and assisting them in better describing their ideas for consideration by the team. Without a skilled facilitator the productivity of the exercises would have been severely limited.

Perhaps the greatest issue was the tendency of the team to seek out procedures (P) and personnel skills and training (PST) or low level engineering controls rather than beginning with controls that could eliminate the hazard (*Figure 38*). In this respect the outcomes of these exercises may not be true Best Practices but the product of a cultural acceptance of relatively high levels of risk. The facilitator expressed that he intruded further into the process than he normally would in attempts to move the group up the hierarchy of control. A commonly repeated phrase "that's just part of mining" was not challenged by the question "why does it have to be?" from the group. In this aspect of risk management it may take some time before the industry learns the relative importance of the hierarchy of control.

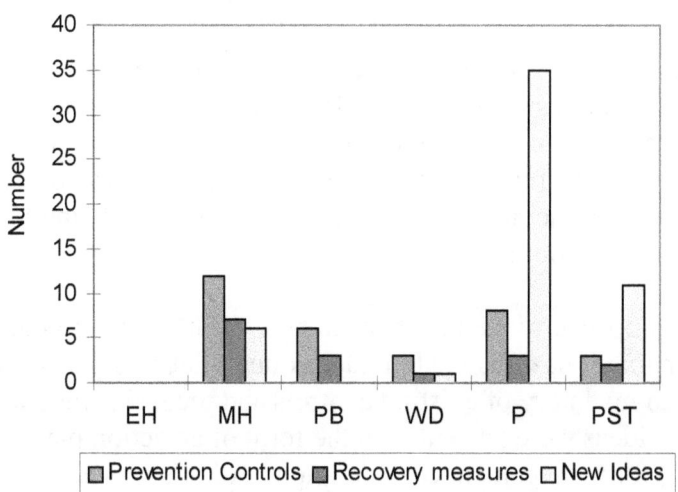

Figure 38 - Distribution of prevention controls and recovery measures for the captive cut-and-fill change of mining method risk assessment.

6.0 – Lessons Leaned

In general, the ten case studies showed that the MHRA process provided information considered beneficial for a safer work environment. The MHRA process can be used to enhance the safety requirements that exist in current regulations or standard operating procedures. This report is not intended to be nor should it be misconstrued as advocating more stringent regulations. It does demonstrate that portions of the mining industry are capable of utilizing the MHRA process and could benefit from its application.

The ten case study examples provide insight as to how the MHRA approach might be used to mitigate the risk from major hazards in US underground mines. Significant threats were identified as well as an inventory of existing controls and recovery measures specific to each threat. Generally new ideas were presented in the form of an action plan (Appendix B) or a risk register (Appendix C) for management consideration. All this was accomplished in a structured, group-oriented activity designed to produce a written report.

6.1 - The Scoping Document

The risk design or scoping document needs to identify an issue of great importance to the mine. These issues were often referred to by the mining personnel as "issues that keep me up at night." As this comment indicates, the risk assessment team should be aware that a frank and open discussion of the hazards is necessary. In the 10 case studies, significant issues were identified and analyzed. However, in one case study (Mine A) the risk assessment team did not feel empowered to address the hazard identified in the scoping document. At case study Mine C, the risk assessment team did not see a compelling need to address the spontaneous combustion threat more than had already occurred at the mine. It is possible that both these issues could have been addressed if the management of the mining operations had clearly identified the hazards under consideration as major threats and communicated its desire to the risk assessment team to find ways to lower the risks associated with these hazards.

6.2 - The Risk Assessment Team

The mines selected their own personnel to participate in their respective MHRA. The makeup and size of each mine's MHRA team was based upon the type and size of the risk assessment topic as shown in *Table 12*. Some teams were quite large as in the case of Mines J and D and some teams were relatively small as in the case of Mine B. The risk assessment team needed the following important characteristics to function effectively: knowledge, diversity, a skilled facilitator, outside experts, training and time.

Knowledge - The case studies demonstrated the need for the risk assessment team to contain key mining operation personnel knowledgeable of the hazard under consideration and familiar with all aspects of the operation. This knowledge should go beyond current regulations and mine practices and should focus on comprehending the root cause of hazards. Innovative solutions depend on this level of knowledge.

Diversity - The diversity of the risk assessment team increased its breath of knowledge and operational perspective. In some cases, the miner familiar with the work process under discussion had a unique perspective that was not always apparent to the professional or management team members. Case study Mines G, H, I and J were observed to have excellent diversity which may have helped the team identify such extensive lists of controls.

A Skilled Facilitator - The facilitator must be well-trained and skilled in the MHRA process. The facilitator is required to always know where the risk assessment team is heading and keeping it on task, and must also deal with dominant personalities and make sure that all voices are heard. It is important for the risk assessment team to have knowledge of the hierarchy of controls concept. It is the facilitator's responsibility to make sure the team has this knowledge.

Outside Experts - The risk assessment team should contain outside experts that have expertise beyond that contained at the mine site concerning the hazard under consideration. Typically outside experts are external to the normal decision-making group at the mining operation. They can consist of technical representatives from manufacturers, consultants familiar with the mining operation, or content experts from academia and government.

Training - The risk assessment team must have a working knowledge of risk assessment tools and techniques. This can be accomplished prior to or during the actual risk assessment. However, time used for training should not limit the time needed to conduct the MHRA.

Time - The risk assessment team must have sufficient time to adequately address its tasks. Case study Mines A and C did not have sufficient time to adequately perform the MHRA exercise. Part of this was due to the need to provide the risk assessment team members with some fundamental training in the basic concepts of risk management.

6.3 – Important Risk Assessment Tools and Techniques

Risk assessment tools and techniques were used during several steps in the MHRA exercise. During Step 1, when the major potential hazards were being identified and characterized, flow charts were often used. These flow charts were especially useful in dissecting work processes. In other cases, when examining operational issues covering the mine site, it was necessary to segment the mine in a logical manner.

During Step 2 of the MHRA exercise, when risks were ranked, the WRAC and PHA were used. If the consequences of a potential unwanted event are high, it may not be necessary to risk rank the potential unwanted events using a WRAC or PHA. Regardless of the method used, it was critical for the risk assessment team to develop a complete list of potential unwanted events associated with the major hazard under consideration and determine which deserved the effort of an MHRA.

After the risk ranking process, the team focused on determining important existing prevention controls and recovery measures (Step 3) and identifying new prevention controls and recovery measures (Step 4). In all case studies with the exception of Mine B, a BTA was used. A BTA was required for each potential threat. The BTA was the most used risk assessment technique during this pilot project.

6.4 - The Risk Assessment Team Outputs (Identified Controls)

The main output of the risk assessment team is the existing and new prevention controls and recovery measures that lower the risk associated with the hazards under consideration. In total, 451 controls were listed during the ten case studies. The minimum was 1 (Mine B) and maximum was 103 (Mine I). There is no correlation between the number of controls and the success of the MHRA. At Mine B, one new idea was identified and it eliminated the hazard. No other controls were needed. In general, procedure (P) controls (44% of total) were used the most by the 10 risk assessment teams and hazard elimination (EH) the least (less than 1% of the total controls) (*Figure 39*). The other control categories ranged from 19% for minimize hazards (MH) to 9% for warning devices (WD).

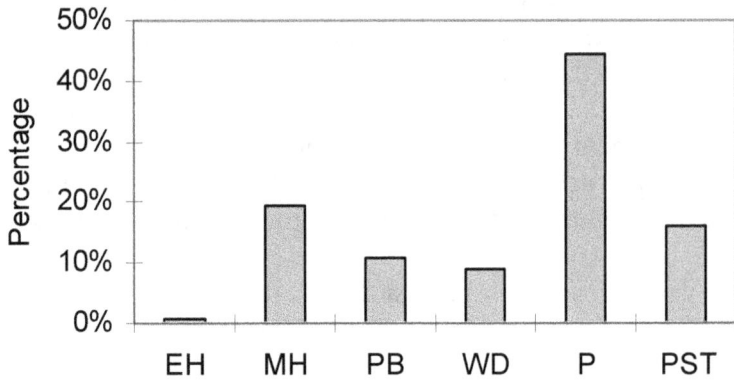

Figure 39 - Percentage of the total controls by category.

The MHRA process requires that hazard elimination be considered as a primary means to reduce risks. In practice, it is often difficult for the risk assessment teams to discuss whether to accept or eliminate the hazard under consideration. This may be partly caused by the team's lack of participants responsible for making these kinds of decisions or partly due to the perceived MHRA objectives. Once a mine is operational, hazard elimination becomes more difficult and is best considered when management specifically requests it and this request is included in the scoping document. A total of three new ideas from three different case studies were classified as hazards elimination. Therefore, seven of the case studies did not consider hazards elimination as a potential control.

MHRA controls rely extensively on the use of regulated standards and company Best Practices. In some cases it was difficult to determine if the risk assessment teams were aware of the differences between existing practices and Best Practice. Outside experts were helpful in identifying controls, barriers or work processes that represented leading industry practices.

6.5 – Documentation

A written document is critical to the MHRA process and, at a minimum, should contain 1) a list of existing prevention controls and recovery measures that can be monitored and audited, and 2) a lists of new ideas for further consideration by management. Management must assign persons to be accountable for existing and newly adopted controls and respond to the team with the reasons for their actions, especially if the assigned persons decide not to act on any recommendation. It is not known if the existing controls were being monitored or audited at the case study mines or if the new ideas were evaluated by management. There was one occasion when management, after reviewing the list of new ideas, asked the risk assessment team to rank their new ideas. The team did not feel that it had enough data or enough time to adequately accomplish this task. There is also a possibility that a consensus would have been difficult to achieve.

7.0 – Success of Risk Assessment Case Studies

A number of key points were recognized from the ten case studies. These key points help to define the degree of success each case study realized in performing an MHRA and provided an opportunity to understand its strength and weaknesses. To help evaluate the degree of success, the MHRA case study exercises were divided into six categories. Each category represents an important aspect of the MHRA exercises. The NIOSH observers compiled information on each case study and made a determination as to how each performed. The six categories are: existing risk management culture, risk assessment design, risk assessment team, risk assessment process, quantity of existing controls, and quality of new ideas. For each of the ten case studies, the demonstrated degree of success was defined in a relative sense as more-than-adequate, adequate, or less-than-adequate (*Table 55*). This evaluation was based on an analysis of the risk assessments team performance and the quality and character of the identified and proposed controls. Three of the ten case studies are rated as performing a more-than-adequate risk assessment, five as adequate, and two as less-than-adequate. A method for self-assessment of mining operations risk management culture is provided in Appendix D.

Table 55 - An assessment of the adequacy / success of the ten MHRA case studies.

Case study mine	Existing risk management culture	Risk assessment design	Risk assessment team	Risk assessment process	Extent of existing controls	Quality of new ideas	Overall
Mine A	L	L	A	NA	NA	NA	L
Mine B	L	A	M	NA	NA	M	A
Mine C	L	A	A	L	L	L	L
Mine D	A	A	A	A	A	A	A
Mine E	L	A	A	A	M	A	A
Mine F	L	A	M	A	A	A	A
Mine G	A	A	A	A	M	A	A
Mine H	A	A	M	M	M	M	M
Mine I	A	A	M	M	M	M	M
Mine J	M	A	M	M	M	M	M

A – Adequate
M – More-than-adequate
L – Less-than-adequate
NA – Not available or did not occur

7.1 – Existing Risk Management Culture

The existing risk management culture of a mining operation impacts the potential for a successful MHRA. One way to measure the suitability of an operational culture for MHRA is to identify the degree to which risk assessment techniques have been previously embraced by the mine. Risk assessment techniques are discussed in Section 3.1 and can be summarized in the order of increasing complexity as 1) informal, 2) basic-formal, and 3) high-level formal.

> Informal risk assessment techniques, i.e. multiple step approaches where workers are asked to look for hazards, determine the significance of the hazard, and take some action to mitigate the risk. Examples include SLAM, Take-Two for Safety, etc. At the case study mines, informal risk assessment techniques were already in use at some mines and not evident at others.

Basic-formal risk assessment techniques, such as Standard Operating Procedures (SOP) and Job Safety Analysis (JSA), establish official procedures and practices for important work practices at the mining operation.

High-level formal risk assessment techniques are structured approaches that incorporate risk analysis tools, such as the HAZOP, FTA, etc., and produce a document that assesses risks. Several mines indicated that these techniques may have been used by the company in the past. Most of the case study mines were not using high-level formal risk assessment techniques.

It is assumed that mines already using informal and basic-formal risk assessment techniques appeared to be in a better position to successfully undertake an MHRA. Undertaking an MHRA could represent a significant change in the way the organization functions. Therefore, it may be more appropriate for mining operations to first begin to implement informal or basic-formal risk assessments before attempting an MHRA. Most case study operations had indicated some use of basic-formal risk assessments techniques in the past, but the quality of these applications was difficult to determine. One clue to recognizing an operation's reliance on these tools was to determine if there was an SOP that defined how other SOPs would be developed, recorded and managed. One case study operation had such an SOP. Of the ten case study mines, five were observed to have either an adequate or more-than-adequate risk assessment operation culture. The other five were assessed as less-than-adequate in that no risk assessment techniques were evident.

7.2 – Risk Assessment Design

The risk assessment design was best executed when the mining operation representatives took MHRA training and prior to the actual MHRA exercise. If training was not possible, extra effort was placed on making sure that as many persons as possible had an opportunity to comment on the risk assessment design document prior to the actual MHRA. The key here is feedback. Remember, the hazard the risk assessment design addresses should be of great importance to the mining organization. The phase "this is the issue that keeps us up at night" was often used by participants to describe the hazard under consideration by the MHRA.

In all but one of the case studies, the risk assessment design was considered adequate. The case study classified as having a less-than-adequate risk assessment design was based on the need to change the design once the MHRA began.

7.3 – Risk Assessment Team

Team selection is a critical component of a successful MHRA. Members of a risk assessment team should be picked carefully, making sure that they are 1) knowledgeable about the hazards under consideration, 2) not overly committed to a particular way of business or work process, and (3) able to express their opinions within a group setting and in the presence of supervisors. All of case study risk assessments were staffed with members that fit these characteristics. However, because most teams struggled with hazards elimination as a potential control, these teams might have benefited from additional persons with the authority and responsibility to address the significant issues associated with hazard elimination.

Diversity is an important aspect of a successful risk assessment team. Solutions to major hazards can come from persons at all levels within a mining operation. This is why labor should be represented as equal participants within the risk assessment team. Both management and labor representatives should be 1) knowledgeable about the hazard under consideration, and 2) familiar with the associated work processes. Team members from labor were usually active in group discussions and very aware of work process details that only come from actually performing the tasks. In several of the case studies, members of the general workforce were key components of the risk assessment team. They also can help to communicate the findings of the MHRA to the general workforce. This was observed to be a powerful tool for implementing the outcomes of the MHRA.

Another desired aspect of a risk assessment team is the use of outside experts, i.e. outside the mining operation. These experts should have special knowledge about the major hazard under consideration that is not currently available on the company's team. This expertise can come from government, academia, manufacturing or consulting agents. These outside experts will help to eliminate attitudes like, "this is the way we have always done it." Five of the risk assessment teams were viewed as more-than-adequate because they used outside experts on their team.

Lastly, training is essential for the risk assessment team members. This can be accomplished prior to the MHRA or it can occur during the MHRA. In fact, every MHRA exercise incorporated training. This must be factored into time needed to complete the MHRA.

7.4 - The Risk Assessment Process

When designing an MHRA, it is important to recognize that the team has flexibility in deviating from the general structure. Very few of the case studies followed every step in the MHRA process (see Section 4.0 for more details). In several cases, the consequence of the event was viewed as significant and no further discussions of event likelihood were warranted (Mines F, H and I). In one case, a thoughtful discussion of an existing work process revealed that a relatively simple change to the process effectively eliminated the hazard (Mine B). This MHRA reached its objective during the first step in its process. In another case, the risk assessment process could not be adequately addressed because the risk assessment design had been altered and there was insufficient time to complete the MHRA (Mine A). In a third case, the one classified as less-than-adequate (*Table 55*), the risk assessment process was abandoned by the risk assessment team (Mine C). Of course, flexibility within the procedure has the potential to produce a wide range of responses. Deviation from a formal risk assessment plan should be under the guidance of a skilled facilitator without a stake in the outcome of the process.

7.5 – The Extent of Existing Controls

An MHRA is undertaken when a mining operation seeks to move beyond simply reacting to standards and regulations and desires to develop a more proactive approach to dealing with its major hazards. The extent of the controls for a particular hazard can be viewed as a measure of the successful application of an MHRA. If the list of controls mirrors that which is required by standards or regulations, then the outcome of the MHRA should be considered less-than-adequate. The list of controls is defined in two ways: those prevention controls and recovery measures currently in place within the mining operation, and the new ideas that help to further mitigate the risks associated with the hazards. The existing controls are the subject of this section, while the new controls are covered in Section 7.6.

The extent of existing prevention controls and recovery measures for the ten case study mines varied widely. Two case studies did not sufficiently analyze to determine their adequacy. One case study was viewed as less-than-adequate from a risk management perspective since the existing controls were completely defined by those required by MSHA regulations. The other seven case studies had either adequate or more-than-adequate existing prevention controls and recovery measures. The more-than-adequate classification was defined by the depth and breath of the controls listed by the risk assessment team.

7.6 – The Quality of New Ideas

One of the most important outputs from an MHRA is the new ideas generated by the team to further mitigate the risk associated with the major hazard under consideration. A risk assessment team should always be expected to produce additional prevention controls and recovery measures beyond the existing mandated standards. That is the main purpose for conducting the MHRA. Failure to produce such results indicates problems with the team composition or the scope of the task. When a team does not search for new ideas, it will fail to suggest actions that will lower risks. Two case studies were unable to produce new ideas – one because it was unable to complete the risk assessment process; the other because it did not feel a need. The other eight case studies produced new ideas that seemed to have merit and deserved to be investigated further by the mining operation.

Distinguishing between an adequate versus more-than-adequate quality of new ideas was accomplished by an assessment of their quantity and character. One way to assess the character of new ideas is to determine their place in the hierarchy of the control (Section 4.4). The most effective controls eliminate or minimize the hazards. Mine B produced one new idea during the initial discussion period that effectively eliminated the hazard. This is without question the most effective form of control. Next, physical barriers are used to separate the worker from the hazard. Less effective controls rely on warning devices requiring manual response and administrative procedures. The least effective controls rely on personnel skills and training that contain many opportunities for human error. The most successful controls emphasize hazard elimination, as was the case for Mine B. An effective MHRA contains a balanced collection of controls. The case studies listed as more-than-adequate (Mines H, I and J) had this balance, while those viewed as adequate relied more on the less effective hierarchy controls.

8.0 – Future Use of the MHRA Process in Mining

In this report, an MHRA methodology was investigated through a NIOSH pilot project that conducted field trials at ten case study mines. The tools and techniques used in the Australian Minerals Industry to conduct an MHRA were reviewed and summarized. Detailed information from the ten case study mines was reported and analyzed. The lessons learned from the case study were examined and the relative success of the MHRA process for mitigating risks was determined.

While this concept is relatively new to the US mining community, it is not new to other US industries with major hazards, i.e., nuclear, petrochemical, aerospace, etc. – nor is it new to many other major mining countries where legislation mandates that sound risk management principles be utilized. Certainly considerable national and international expertise exists today to help those that are interested in becoming more proactive in dealing with their major hazards. These efforts could be strengthened with an accepted framework for conducting an MHRA and facilitated with appropriate training and instructional guidelines and resources.

All mines have the potential for major hazards that can fatally injure miners and threaten the well-being of the mining operations. It therefore seems prudent for all mines to consider MHRA as a means to proactively address safety threats to their mining operations. This is seen as a means of going beyond merely complying with existing mining regulations, by systematically examining hazards capable of producing significant consequences. MHRA is also needed where a significant change is planned to equipment, machinery, procedures, or manner of working.

Mine management needs to lead the planning, organizing, controlling and motivating efforts in support of the MHRA approach. Management must understand that an MHRA can produce a design recommendation but should not attempt to produce the actual engineering design. In most cases, the teams did not have the right make-up for engineering design work. Suppliers of goods and services and regulatory agencies are also an important partner and can assist in the risk assessment team. Labor should be given the opportunity to participate in the process and take an active role in implementing change. One means of evaluating if an organization and its management are ready for the MHRA process is through a self-assessment of its current management practices. Appendix D provides a means for an organization to identify how it addresses risk and hazards in a qualitative format. MHRA is best applied in organizations that have moved beyond reactive management of hazards and have mastered basic-formal risk assessment processes.

An MHRA can be very effective if used early in a project's life cycle, when systems and work processes are being designed, and should be considered in conjunction with other risk assessment activities, i.e. financial, environmental, etc. Treating a hazard early in the life cycle, when systems are being designed and equipment specified, can be done more efficiently and effectively than later in the cycle, when most work processes revolve around maintaining existing designs.

It is critical that the risk assessment be designed to exploit the strengths of the MHRA approach and to avoid its weaknesses. The strengths of the MHRA approach are its ability to

1. set clear direction to solve specific high-risk problems,
2. focus on priority concerns,
3. get involvement and commitment from a wide cross-section of the mine's work force,
4. decrease potential losses for a mining operations,
5. help to build teams to solve major mining issues,
6. go beyond merely complying with existing mining standards and regulations, and
7. focus upper management attention on issues existing at the operational level (this is where the written documentation can be very helpful).

The weaknesses or threats of the MHRA approach are its need to
1. focus on changes within the existing way the mine conducts business,
2. take time away from activities directly related to production,
3. put additional time constraints on a mining operation's "best people,"
4. introduce the cost of implementing new prevention controls and recovery measures,
5. potentially alter a mining operation's priorities,
6. need for there to be an existing risk management structure to build upon, and
7. need for an openness in management / labor communications.

In general, the ten case studies demonstrate that most US mines have the capability to successfully implement an MHRA and that the MHRA methodology produced additional prevention controls and recovery measures to lessen the risk associated with a select population of major mining hazards. The basic ingredient for a successful MHRA is the desire to become more proactive in dealing with the risks associated with events that can cause multiple fatalities. If a mining operation does not commit sufficient time or is not willing to utilize its most experienced personnel to this effort, it is unlikely to produce a successful outcome. A successful outcome is marked by a thorough examination of existing prevention controls and recovery measures and the generation of new ideas to further mitigate the risks associated with the major hazards under consideration. It is also essential for a mining operation to be receptive to a written report that discusses key existing controls and identifies who will investigate the risk assessment team's list of new ideas.

All mines arguably have major hazards with the potential to produce multiple fatality events. The ten case studies demonstrate that most mines currently go beyond the minimum requirements for mitigating the risks associated with major hazards. When pressed to consider more controls to further mitigate the risk, a well-staffed risk assessment team was able to identify additional controls. For these mining operations, it was important to add additional controls, even if they were not required by existing mining regulations, to lower the risks associated with the major hazards under consideration. These operations realize that the negative consequences associated with these potential unwanted events require additional actions and that current regulations do not totally protect their miners from all the specialized site-specific hazards shaped by local characteristics. The MHRA methodology represents a structured approach that helps mining operations develop additional controls aimed at mitigating the risk associated with their most significant hazards.

While all mines have major hazards, not all mines are prepared to utilize an MHRA. If a mining operation is not willing to commit its best people to an MHRA or will not provide them with sufficient time to see the process through to its conclusion, the MHRA output may prove to be useless. Additionally, if a mining operation is not prepared to discuss its major hazards in an open and honest fashion and to present the findings of the risk assessment in a written report, the MHRA output will be unclear, and attempts to monitor or audit important controls may not be possible.

An MHRA can be most effective when the mining operation possesses 1) a proper understanding of its hazards, 2) experience with informal and basic-formal risk assessment techniques, 3) proper facilities, machinery and equipment, 4) suitable systems and procedures that represent industry Best Practice, 5) appropriate organizational support with adequate staff, communications and training, 6) a formal and thorough plan for emergency response, and 7) a safety risk management approach that is promoted and supported at all levels of the organization.

9.0 – References

Anon, 2005, "An Unplanned Detonation of a Blast Hole Occurred at a Surface Coal Mine in Indiana," Mine Safety and Health Administration, Safety Hazard Alerts webpage, June 10, 2005.

CFR, 2005, "Escapeways," Code of Federal Regulations, Part 30, Section 57.11050.

CMEWA, 2003, "Review of the Mine Safety Inspection Act of 1994," Chamber of Miners and Energy, Western Australia.

Freeman, M. "Observations on Mine Safety Management from Review of Major OHS Prosecutions and Investigations", NSW Department of Primary Industries, Sydney, 2007

Gates, R.A., R.L. Phillips, J.E. Urosek, C.R. Stephan, R.T. Stoltz, D.J Swentosky, G.W. Harris, J.R. O'Donnell, Jr., and R.A. Dresch, 2007, "Fatal Underground Coal Mine Explosion, January 2, 2006, Sago Mine, Wolf Run Mining Company, Tallmansville, Upshur County, West Virginia," Report of Investigation, Mine Safety and Health Administration, May 9, 2007, 189 p.

Grayson, L., A. Bumbico, S. Cohn, A. Donahue, J. Harvey, J. Kohler, T. Novak, C. Roberts, and H. Webb, 2006, "Improving Mine Safety Technology and Training: Establishing U.S. Global Leadership," Mine Safety Technology and Training Commission, National Mining Association, 193 p.

Holmberg, R. and D. Salomonsson, 2002, "Snap, Slap and Shoot – A Possible Cause for Premature Ignition of Shock Tube," International Society of Explosive Engineers, 2002 General Proceedings Collection - Volume II, pp. 91-103.

Hopkins, A., 2000, A Culture of Denial: Sociological Similarities between the Moura and Gretley Mine Disasters," Journal of Occupational Health and Safety Australia and New Zealand, Vol. 16, No. 1, 2000, pp. 29.36.

Iannacchione, A.T., G.S. Esterhuizen, S. Schilling, and T. Goodwin 2006, Field Verification of the Roof Fall Risk Index: A Method to Assess Strata Conditions, Proceedings of the 25th International Conference on Ground Control in Mining, Morgantown, WV, Aug., pp.128-137.

Iannacchione, A.T., G.S. Esterhuizen, L.J. Prosser, and T.S. Bajpayee, 2007, Technique to Assess Hazards in Underground Stone Mines: The Roof Fall Risk Index (RFRI), Mining Engineering, Vol. 59, No. 1, pp. 49-57.

Jobs, B., 1987, Inrushes at British Collieries: 1851 to 1970, Colliery Guardian, Vol. 235, No. 5 and 6, May and June, pp. 192-9 and 232-5.

Joy, J., 2006, Minerals Industry Risk Management Framework, Minerals Industry Safety and Health Centre, University of Queensland, 83 p.

McAteer, D., 2007, Testimony Before the Committee on Education and Labor, United States House of Representatives, April 21, 2007.

Moebs N.N. and G.P. Sames, 1989, Leakage Across a Bituminous Coal Mine Barrier, USBM RI 9280, 17 p.

NSWDPI, 1997, Risk Management Handbook for the Mining Industry: How to conduct a risk assessment of mine operations and equipment and how to manage risk, New South Wales Department of Primary Industries, MDG 1010, May 1997, 95 p.

QDME, 1998, Recognised Standard for Mine Safety Management Systems, Queensland Department of Mines and Energy, Safety and Health Division, Coal Operations Branch, January, 1999, 6 p.

QMC, 1999, Information Paper – Safety and Health Management for Queensland Mines and Quarries, Queensland Department of Mines and Energy and the Queensland Mining Council, Brisbane, Australia, 25 p.

Robertson, A. MacG. and S. Shaw, 2003, Risk Management for Major Geotechnical Structures on Mines, in Proceedings of Computer Application in the Mineral Industry, CAMI, Calgary, Alberta, CA, Sept. 8-10, 2003, 18 p.

Smith, A.C, W.P. Diamond, T.P. Mucho and J.A. Organiscak, 1994, Bleederless Ventilation Systems as a Spontaneous Combustion Control Measure in U.S. Coal Mines, US Bureau of Mines IC 9377, pp. 1-43.

Standards Australia, 2004, Risk Management, AS/NZS 4360:2004, Sydney, Australia, 26 p.

Watzman, B, 2007, On Protecting the Health and Safety of America's Mine Workers Testimony of Bruce Watzman, Vice President of Safety and Health, National Mining Association, Before the Committee on Education and Labor, United States House of Representatives, March 28, 2007, 6 p.

APPENDIX A – Example of a Bow Tie Analysis (BTA).

In the paper, parts of a BTA were often provided to demonstrate the application of this risk analysis technique. *Table 56* and *Table 57* are exhibits of a fully completed BTA.

Table 56 - Left side BTA for Mine I.

Causes	Prevention Control Measures	
Top Event = Fire on Locomotive or mantrip		
1 - Short circuit	PC1	Maintenance, weekly checks of loco (general inspection) *[P]*
	PC2	Pre-operation check by operator (brakes, trams, etc.) *[P]*
	PC3	Operator is experienced enough to recognize abnormal operation *[PST]*
	PC4	Locomotives have fuses and breakers *[MH]*
	PC5	Some cables are protected in conduits *[PB]*
	PC6	Operators training *[PST]*
	PC7	Radio communication is available to get assistance/advice, re abnormal operations *[PST]*
	PC8	Abnormal operation is reported to supervisor and maintenance shop *[P]*
	PC9	Locomotive designed to standard *[MH]*
	PC10	State electrical inspector inspects any loco that has had major rebuild *[P]*
	PC11	Mine personnel inspect new equipment and rebuild before it is used *[P]*
2 – Overheated battery (water, load, charge)	PC12	Maintenance batter is cleaned, watered, checked for dead cells as part of a battery maintenance program (contractors and others) *[P]*
	PC13	Supervisors know motor operator and they select competent operators for heavy/difficult loads *[PST]*
	PC14	Pre-operation check by operator – battery OK (water levels, clean, etc.) *[P]*
	PC15	Operator is experienced enough to recognize abnormal operations (low power) *[PST]*
	PC16	Battery is charged when trolley wires and gage indicate charge level *[WD]*
	NI1	Reinforce and follow the requirements of the maintenance program or batteries, consider checklists/verification procedure that it is being followed *[P]*
	NI2	Investigate defining a specific percentage battery charge that is minimum to enter panel *[P]*
	NI3	Add checking gage accuracy in the battery maintenance program *[WD]*
	NI4	Investigate whether there is an identifiable level of complexity/experience for major loads, thereby a list of heavy load operators *[P]*
3 – Battery short circuit	PC17	Maintenance battery is cleaned, watered, checked for dead cells as part of a battery maintenance program (contractors and others) *[P]*
	PC18	Pre-operation check by operator – battery OK/ obvious damage *[P]*
	PC19	Safe Work Instruction and Best Practice Teams sometimes observe pre-op inspections on Locomotives *[P]*
	PC20	Operator is trained to open breakers (2) and take plugs off batteries if there is a short. If that doesn't work, main lead is disconnected *[P]*
	PC21	Battery rebuilds are done to mine specification *[P]*
	As above: see *PC4 and PC9*	
	NI5	Reinforce the need to remove any baking soda that has been used to absorb water or batteries so that it doesn't become a conductor (during pre-operation inspection) *[P]*
4 – Resistor overheating and trash on loco hot spots burns	PC22	Resistors are designed to standards and government inspected *[MH]*
	PC23	Additional fans installed on some locomotives to provide additional cooling *[MH]*
	PC24	Operator is aware of hot resistors because braking will be reduced and other locomotive is then used to brake *[PST]*
	PC25	Operator stops machinery and lets resistors cool if overheated (breaking reduced) *[PST]*
	PC26	Resistors are closed in so trash in area is unlikely *[MH]*
	NI6	Add checking inside the resistor area (lift lid) for any combustibles to pre-operation inspection (NOTE that dust can get into resistor compartment) *[P]*

	NI7	Investigate changing or modifying resistors to perform under load without overheating *[MH]*
5 - Brake lockup, left on or applied due to bleed off, causing failure and brake fluid/grease fire (mantrip)	PC27	Design to standard/inspected *[MH]*
	PC28	Pro-operation test brakes before operating equipment *[P]*
	PC29	Maintenance – brakes are fixed as required *[P]*
	PC30	Normal braking is electrical so overheating of other brake is unlikely *[MH]*
	PC31	Operator smells brake heat if park brake left on *[PST]*
	PC32	There is a red light to indicate brakes are on *[WD]*
6 - Fire occurs with explosive/oil on locomotive	PC33	Loads are about 10 ft from locomotive heat sources (hydraulic oil/wood equal fuels) *[PB]*
	PC34	Any explosives are transported in specialized container and hauled separately *[PB]*
Top Event = Fire with High-Voltage System		
1 - HV line hit by derail/impact form load	PC35	Rail maintenance program where each shift is given an area of track to install and maintain (to standards for track installation) *[P]*
	PC36	There is a pre-shift examination by the mine examiner *[P]*
	PC37	Locomotive operators report any track issues to supervisor *[P]*
	PC38	Load guidelines are applied at the mine to avoid oversized/shifting loads that might derail the locomotive *[P]*
	PC39	HV shielded cable, located in rib/roof corner reduces likelihood of damage, there is also some guarding *[PB]*
	PC40	HV circuit breakers/GFCI/pilot circuits protect system from overload/fault/fires and are tested and recorded on a monthly basis *[MH]*
	PC41	HV hung to regulatory requirements *[P]*
	PC42	Examiner inspects the area including HV *[P]*
Top Event = Fire due to welding/cutting		
1 - Trash/loose coal in area ignited by welding/cutting	PC43	Each shift is responsible for cleaning up trash in an area *[P]*
	PC44	Trash is bagged and put on empty car *[P]*
	PC45	Locomotive operators pick up trash on outby areas around tracks *[P]*
2 - Damaged/faulty torches or hoses lead to fire (practices)	PC46	Mine has strict procedures for cutting and welding underground *[P]*
	PC47	Qualified person must be present to make CH_4 checks including checking area before they leave *[P]*
	PC48	Fire protection and rock dust is included in the procedure *[P]*
	PC49	Where possible a water line (charged) is also taken to the area *[P]*
3 - CH_4 blowers ignited by welding/cutting	As above: see PC49 about qualified person and CH_4 monitoring	
Top Event = Fire on rock duster battery car		
1 - Rock duster battery packs battery shorts/faults	PC50	Rock dusters are designed to standard and inspected before underground use *[MH]*
	PC51	Dedicated crew takes care of charging and inspecting equipment, including operator being in the area during operation *[P]*
	PC52	There is a weekly electrical check of the stone dusting equipment by maintenance *[P]*
2 - Compressor overheats	PC53	Compressors are designed to standard and inspected before underground use *[MH]*
	PC54	There is a weekly electrical check of the compressor equipment by maintenance *[P]*
	PC55	Operators check oil in compressors *[P]*

PC – Prevention Controls
NI – New Ideas
MH – Minimize Hazard
PB – Physical Barrier
WD – Warning Devices
P – Procedures
PST – Personnel Skills and Training

Table 57 - Right side BTA for Mine I.

Consequences	Recovery Measures	
Top Event = Fire starts on the Locomotive/mantrip		
Small fire becomes big fire (lost assets)	RM1	All locomotives are fitted with heat sensors and manually initiated fixed fire suppression and hand-helds (20 lbs) *[WD]*
	RM2	Hand helds are located so they can be easily accessed in a fire at battery/brakes/etc. *[P]*
	RM3	Operators are trained re: hand-held/charged system every 2 years *[PST]*
	RM4	Fixed and hand-helds are checked regularly *[P]*
	RM5	Pre-operation checks including making sure hand-held is charged and pinned *[P]*
	RM6	Persons are trained that an air line can be charged to a two-inch water line to provide fire fighting water to the track heading *[PST]*
	RM7	There is also a return water line that can supply water to fight a fire until air pressure to face is lost *[MH]*
	RM8	There is a real time continuously monitoring system that detects CO located every 2500 feet in track heading *[WD]*
	RM9	System is linked to the underground bunker and outside surface hoist house that is continuously monitored by a person *[P]*
	RM10	System alarms at 5 ppm CO (alert) + 10 ppm CO (alarm). System also has a malfunction alarm *[WD]*
	RM11	Person reacts to alarms by notifying shift foreman and persons in area using radio or telephone. That person goes to check area with CO detector. *[P]*
	RM12	There is a CO alarm at the conveyor tail (10 ppm CO) in the panel *[WD]*
	RM13	The mine has a designated Responsible Persons (RP) who is notified if a fire (small) is identified and appropriate other notifications are formalized *[P]*
	RM14	All persons evacuate the mine if a big fire is identified *[P]*
	RM15	RP makes decisions about actions to be taken underground to fight fire, change ventilation, etc. *[P]*
	NI8	Make operators aware that, if possible, when there is a small fire or smoke from a locomotive/mantrip/rock duster there may be an opportunity to reduce/stop smoke to the face by putting equipment into a switch/spur track and open man-door into the return heading to short circuit into the return *[P]*
Top Event = Fire caused by high-voltage		
Small fire becomes big fire (lost assets)	RM16	Jackets and insulation are fire resistant *[PB]*
	RM17	System is designed to de-energize quickly (GFCI, etc.) *[MH]*
	RM18	HV cable hung in a manner (location) so it is not exposed to materials that can come from a short, high-temperature heat source *[P]*
	As above: see *RM6 to RM15*	
Top Event = Rock duster/compressor fire		
A - Small fire becomes big fire (lost assets)	RM19	Equipment has fixed, automatic and manually operated fire suppression on the battery areas and mounted hand-helds *[PB]*
	RM20	Operators are trained every 2 years *[PST]*
	RM21	Certified contractor does maintenance inspection on the fire suppression system every 6 months *[P]*
	RM22	There are weekly inspections to see if system is faulted and if it alarms if faulted or discharged *[P]*
	As above: see *RM6 to RM15*	
	NI9	Investigate whether fixed fire suppression can be located over/at compressor *[PB]*
Top Event = Cutting/welding fire		
Small fire becomes big fire (lost assets)	RM23	Hand-helds are part of welding equipment used underground *[P]*
	RM24	Area will be rock dusted before welding *[MH]*
	As above: see *RM6 to RM15*	
	NI10	Put one joint of fire hose to be carried on the track jeep at all times *[P]*

Top Event = BIG FIRE		
Miners affected by smoke at the face	See previous re: phones, conveyor shut down warning, CO alarms, visible smoke/smell re: warn of fire	
	RM25	Supervisor maybe present with portable gas monitor to deal with CO level [WD]
	RM26	Persons on face are trained to put on M20* self-rescuers, leave the panel if dense smoke is in intake, meet at the power center, grab an extra SCSR, tag together, go to return, and use lifeline in return to egress the section (M20 O_2 20-minute supply). [P]
	RM27	There is a cache of 1 hour SCSRs at the load center on mobile equipment and caches are located 5700 feet in the intake track entry and 5700 feet in the return (staggered every 2850 feet down panel) [PB]
	RM28	Lifelines lead to caches and there are two cones on line to alert that a door or SCSR cache is present [P]
	RM29	There is a practice return egress escape every quarter (intake twice per year and return twice per year) [P]
	RM30	Caches are located in cross-cuts with doors in stoppings [PB]
	RM31	Persons are trained to take an extra SCSR [PST]
	RM32	Per MSHA requirements barricading materials have been located in panels and persons have been made familiar with methods of building barricades (escape, escape, escape...) [PB]
	RM33	There are trained and qualified mine rescues teams available to attempt underground rescue [PST]
	RM34	Mine ER Plan includes external and internal communication, external medical services, family notification, security, etc. [P]
	NI11	Add clarification to ER training, re: egress in light smoke – i.e. if light smoke in intake use transportation to exit as far as possible* then cross to return to egress (* smoke is too dense to see ahead) [P]
	NI12	Make sure the caches are located in cross-cuts with doors in stoppings [P]
	NI13	Reinforce the need to put self-rescuer or, if closer by, SCSR on as soon as any smoke is detected (issue: smoke may get worse and easier/safer to don SCSR now) [PST]
	NI14	A method should be developed to access stopping doors at caches to check if intake is fresh air so that a person can remain attached to lifeline and/or team. The method should be included in 90-day ER training. [P]

RM – Recovery Measures
NI – New Ideas
MH – Minimize Hazard
PB – Physical Barrier
WD – Warning Devices
P – Procedures
PST – Personnel Skills and Training

APPENDIX B – Action Plan of New Ideas.

	Identified Potential New Controls	Specific Required Actions	Responsibility	Due date
Design	Investigate changing or modifying loco resistors to perform under load without overheating			
Design	Investigate whether fixed fire suppression can be located over/at compressor			
Design	Put one joint of fire hose on loco to be carried on the track jeep at all times			
Maintenance	Reinforce and follow the requirements of the maintenance program for batteries, consider that checklists/verification procedures are being followed			
Maintenance	Add checking gauge accuracy in the battery maintenance program			
Operations	Investigate defining a specific percentage battery charge that is minimum to enter panel			
Operations	Investigate whether there is an identifiable level of complexity/experience for major loads, thereby creating a list of heavy load operators			
Operations	Reinforce the need, during pre-operation inspection, to remove any baking soda that has been used to absorb water or batteries so that it doesn't become a conductor			
Operations	Add checking inside the loco resistor area (lift lid) for any combustibles to pre-operation inspections			
Operations	Make operators aware that, if possible, when there is a small fire or smoke from a loco/mantrip/rock duster there may be an opportunity to reduce/stop smoke to the face by putting equipment into a switch/spur track and open man-door into the return heading to short circuit into the return			
Fire and Emergency Response	Add clarification to ER training, re: egress in light smoke – i.e. if light smoke in intake use transportation to exit as far as possible* then cross to return to egress [* smoke is too dense to see ahead]			
Fire and Emergency Response	Make sure the caches are located in cross-cuts with doors in stoppings			
Fire and Emergency Response	Reinforce the need to put self-rescuer or, if closer by, SCSR on as soon as any smoke is detected (issue: smoke may get worse and easier/safer to don SCSR now)			
Fire and Emergency Response	A method should be developed to access stopping doors at caches to check if intake is fresh air so that a person can remain attached to lifeline and/or team. The method should be included in 90-day ER training.			

APPENDIX C – Risk Register

Hazard/Risk Register Template[9]

Project No:		Section of Facility:		Date:		Page:
Description of Scenario:				Team Leader:		
				Team Members:		
Reference Documents:						
				Minutes By:		

Item No	Initiating Event	Description of Potential Consequences (including magnitude and Effects)					Existing Control Measures					Description of Likelihood of Potential Effects (On/off site) and Likelihood Rating	Risk Ranking	Actions
		Type And Magnitude	Description of Potential Effects (on site and off) and consequence Rating				Description	Critical Control?	SMS Ref	Performance Std NO	COP Data Sheet			
			People	Biophysical Environment	Property	Economic Impact								

---

[9] Adapted from MIHAP No 3 Planning NSW Hazard Identification, Risk Assessment and Risk Control

APPENDIX D - Risk Management Culture and Self-Assessment

Organizational Progress in Risk Management

A number of leading authors on the subject of risk management culture have identified the changing nature of an organization as it progresses along its journey of embracing the principles of managed risk. Hudson, Clemmer and Joy have all discussed a 5-stage model to identify an organization's starting point and gauge the degree of change necessary to effectively implement higher order risk management principles. Measuring where an organization fits in this model is a subjective exercise that evaluates a variety of management systems for their relative degree of resiliency.

The lowest rung on the ladder or first step in the journey is an organization described as vulnerable, non-caring or pathological. Such an organization would view the minimum legal requirements as a significant burden and would be unlikely to pursue formal risk management systems due to the additional obligations of time and resources required. On the other end of the spectrum, the highest order organization is described as resilient, fully integrated, or generative. None of the leading practitioners of risk management feel they have yet reached this level as an organization. The middle of the journey or ladder is the zone where most mining operations find themselves.

A Self-Assessment Tool

The following self-evaluation is intended to provide mine managers with a tool to identify the strengths and weaknesses in their organization and changes needed for successful implementation of an MHRA.

For each category select the description that best fits your mine:

Informal Risk Management Systems:
1. Workers inspect their own workplace, correct as needed, no documentation unless required by law
2. Workers inspect their own workplace, correct as needed, document findings, records retained for review
3. Workers inspect their own workplace, correct as needed, document findings, records retained for review, supervisors randomly audit during shift
4. Workers inspect their own workplace, correct as needed, document findings, records retained for review, supervisors audit during shift
5. Workers inspect their own workplace, correct as needed, document findings, supervisors audit during shift, records reviewed by mine management and follow-up actions pursued.

Task/Job Training Systems:
1. Peer to peer (Hands-On) task training, no formal documented work practices
2. Peer to peer training based on written work practices that are updated as needed
3. Formal training program using dedicated trainers and written work practices that are updated as needed
4. Formal training program using dedicated trainers and written work practices that are reviewed regularly and the performance of the trainers routinely audited. Mine management assigns responsibility for action items
5. Formal training program using dedicated trainers and written work practices that are reviewed regularly and the performance of the trainers routinely audited by local and corporate safety departments. Mine management reports on action items to corporate safety management

Accident Investigations and Analysis:
1. The front line or safety supervisor investigates reportable accidents and gives the reports to the safety department for processing
2. The front line or safety supervisor investigates all accidents, including near misses and gives the reports to the safety department for processing
3. A team investigates all reportable accidents and provides a report to management
4. A team investigates all accidents including near misses and provides a report to management. Management assigns responsibility for action items
5. A team investigates all accidents including near misses and provides a report to management. Management assigns responsibility for action items.

Hazard (Fire, Gasses, Dusts, Ground) Monitoring Systems:
1. Monitoring systems meet minimum legal requirements, reporting as required by law
2. Mostly manual monitoring systems of common hazards with information recorded as required. Action levels require reporting to Supervisors for response decisions
3. Mix of automatic and manual monitoring systems with information recorded and monitored by supervisors in line with formal response procedures
4. Mostly automatic monitoring systems with real time monitoring by supervision and pre-determined automatic responses. System performance audited routinely
5. Mostly automatic monitoring systems with real time monitoring by supervision and pre-determined automatic responses. System performance audited routinely. Parameters monitored include potential but not yet regulated hazards to health and safety.

Monitoring of Behavior:
1. No monitoring of behavior with respect to safety is done other than what is required by law
2. Informal behavior monitoring programs have been introduced on some topics. Review of compliance occurs during incident/accident investigations
3. Some behavior monitoring occurs through formal systems and training on selected topics. Investigations of incidents and accidents compare behaviors to expectations
4. Expectations on behavior are well-defined, communicated, formally monitored and periodically reviewed on priority topics. Feedback on expectations and compliance obstacles from the workforce is sought and incorporated into expectations
5. Expectations on behavior are well-defined, communicated, formally monitored and periodically reviewed by all workers on all topics. All levels of the workforce serve as monitors and enforcers of expectations.

Auditing of Expectations:
1. No auditing of safety performance occur other than by MSHA or State inspectors
2. Auditing is part of accident/incident investigations
3. The safety department conducts a periodic audit for compliance with the rules, MSHA and internal
4. Auditing is done by all levels of management to check for compliance with existing rules and to identify priority issues
5. A formal site-wide monitoring system is in place to both review practices and verify work systems are performing as identified in site-specific risk assessments. The site-wide systems are periodically audited by others.

Culture:
1. We do what we have to, sometimes we get caught
2. We do the best we can, discipline those who don't
3. We put every effort into complying with the law every day. We change systems when we have trouble with compliance. We get few citations
4. The front line managers lead the safety department in identifying problems and solutions. MSHA isn't much of a problem because we always do more
5. All levels of the organization promote safety as the first priority. Regulators look at us as an example of how to manage hazards. Proposed new rules are already a way of doing business here.

131

Evaluation and Paths to Progress:
Add the numbers of each selected response for your score to determine where you stand with suggestions on paths to improvement:

7-10 <u>Vulnerable:</u> Accept the need for change. Embrace informal risk assessment tools. Follow accidents and incidents with investigations that produce JSAs to prevent a reoccurrence.

11-17 <u>Reactive:</u> Understand human error and accept that most of it is unintentional. Formalize standard procedures and minimize reliance on administrative controls. Expand investigations beyond serious incidents.

18-24 <u>Compliance Driven:</u> Emphasize quality and competency of training over time. Pursue formal risk assessments on topics which have not yet been a problem. Increase internal monitoring to ensure conformance. Increase consulting within all levels of the workforce for acceptable solutions.

26-31 <u>Proactive:</u> Fully integrate risk management into all decision systems and projects. Shift focus on eliminating rather than managing hazards. Drive controls up the hierarchy. Open communications within the organization. Seek out 3rd-party auditing.

32- 35 <u>Resilient:</u> You apparently don't need this document.

The steps along this path are like the steps along a journey. This journey is one of building a strong culture of safety (*Figure 40*).

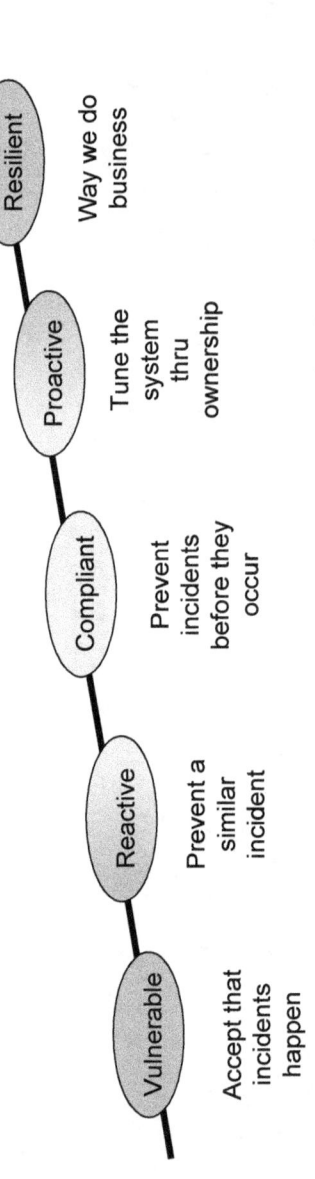

Figure 40 - Steps along the path to an improved safety culture.

www.ingramcontent.com/pod-product-compliance
Lightning Source LLC
Chambersburg PA
CBHW080256180526
45167CB00006B/2551